Springer Series in Operations Research and Financial Engineering

Series Editors:
Thomas V. Mikosch
Sidney I. Resnick
Stephen M. Robinson

For further volumes:
http://www.springer.com/series/3182

András Prékopa • János Mayer
Beáta Strazicky • István Deák • János Hoffer
Ágoston Németh • Béla Potecz

Scheduling of Power Generation

A Large-Scale Mixed-Variable Model

 Springer

András Prékopa
Department of Statistics
Rutgers University
Piscataway, NJ, USA

János Mayer
Department of Business Adminstration
University of Zurich
Zürich, Switzerland

Beáta Strazicky
Budapest, Hungary

István Deák
Department of Computer Science
Corvinus University of Budapest
Budapest, Hungary

János Hoffer
IT Quality Assurance Section
Allianz Hungária Insurance Company
Budapest, Hungary

Ágoston Németh
Ex-Lh Ltd.
Budapest, Hungary

Béla Potecz
(deceased)

ISSN 1431-8598   ISSN 2197-1773 (electronic)
ISBN 978-3-319-07814-4 ISBN 978-3-319-07815-1 (eBook)
DOI 10.1007/978-3-319-07815-1
Springer Cham Heidelberg New York Dordrecht London

Library of Congress Control Number: 2014943241

Mathematics Subject Classification (2010): 90B06, 90B10, 90B30, 90B35, 90C05, 90C06, 90C11, 90C20, 90C30, 78A55, 81V99, 91B74, 94C99

Printed on acid-free paper

Springer is part of Springer Science+Business Media (www.springer.com)

Preface

This project grew out of the 1973 master's thesis that Ágoston Németh, a student of András Prékopa at the Technical University of Budapest, wrote on the problem of optimal power dispatch from producers to consumers. It was read by Béla Potecz, Deputy Director of the Hungarian National Power Dispatch Center, and the three of us decided to create a suitable model for optimal daily power generation in Hungary. We were able to secure financial support from the National Power Company and launched the project at the Computer and Automation Institute of the Hungarian Academy of Sciences, where András Prékopa served as Director of the Department of Operations Research (which in 1977 became part of the larger Department of Applied Mathematics under the leadership of András Prékopa).

The first model with a solution algorithm and computer code was ready in 1979, but it was a failure. It ran successfully on a small network of ten nodes, but the Hungarian power network had 170 nodes. The data also contained many inaccuracies, and the solutions to load flow subproblems were slow and required enhancements.

Our ambition was to handle simultaneously, in one model, the unit commitment and the distribution problem in a given network, taking into account the physics of the transmission network. This resulted in a large-scale, nonlinear, mixed-variable, decomposition-type optimization problem whose solution was still unrealistic given the state of computer technology in the late 1970s and early 1980s. Size reduction was needed, which resulted in the introduction of the concept of *mode of operation*. This meant grouping the generators, and those in one group were supposed to be in operation or in standstill position simultaneously. The problem of the modified model was solved by subsequent uses of a modified Benders decomposition on an IBM 3031 computer. The solution that provided us with the optimal daily scheduling of power generation in Hungary took only 2 min. A day was subdivided into 26 periods, and a very accurate power demand forecast, developed separately by another team in the same Institute, was used.

Many years have passed since the first successful solution, and the methodology from that period was used by the National Power Dispatch Center for some time.

Currently the power plants have many owners, and hence application of the model is difficult, but negotiations are under way regarding its use or modification. We are convinced that the summary of our project presented in this book can still be useful.

Our model belongs to the class of security-constrained unit commitment (SCUC) models that provide an extension of traditional unit commitment models by incorporating security constraints with respect to power flow along a transmission network. A bibliographical survey for the 35 years up to 2003 [54] showed that the first SCUC model that included constraints on the voltages at the nodes of the transmission network was proposed and numerically tested by Ma and Shahidehpour [47]. This model contained two separate subproblems for real and reactive power flow constraints. A simplified model with a single subproblem was presented in [24]. From an algorithmic point of view, the main idea is to apply Benders' decomposition with subproblems corresponding to the power flow component. For large-scale power systems, further developments related to the SCUC model with voltage constraints, as well as the application of the Benders decomposition algorithm, can be found in [49], [23], and [73, 74]. Regarding solution algorithms for the optimal power flow problem, we refer the reader to [21, 22] for extensive bibliographical surveys.

A distinguishing feature of our SCUC model is that in addition to production, startup, shutdown, and changeover costs, a term representing transmission losses is also included in the objective function, and this term is present in the supply constraints as well. The majority of papers in this field focus on either a power systems engineering or operations research approach. A second distinguishing feature of our book is that it combines insights from power systems engineering and operations research, both for building a model and for the development of a solution algorithm. By providing sufficient details, we aim to make the book accessible to readers from both fields and to graduate students.

In this book we assume that the transmission network topology does not change across the scheduling period. However, system reliability and performance can be improved by switching transmission lines on or off. Recent research has suggested that network topology, in connection with the availability of transmission lines, and power generation should be optimized simultaneously; see [53] and references therein. In emergency situations, or to avoid such situations, it may be necessary or advisable to split the transmission network into self-sufficient subnetworks, called *islands*. In [20] the authors propose a mixed-variable model for the optimal formation of such islands.

The subject of our book is short-term power generation scheduling, with the goal of operating existing generating and electric apparatus as a whole at an optimal level. For long-term power system planning, see the survey paper [25].

We do not intend to include all details acquired in the course of the project but rather concentrate on the developed mathematical model and its numerical solution. What follows is a brief summary of the contents of the book.

In Chap. 1 we summarize the most important knowledge concerning electric power systems and formulate the problem from a physical point of view.

In Chap. 2 we disregard the special properties of the Hungarian power system and formulate a general model for scheduling daily power generation by thermal power plants and transmitting power to consumers through a given transmission system. Integer variables represent modes of operation, and constraints representing the network are included.

In Chap. 3 simplifying hypotheses are introduced. They play an important role in the specialized problem, the optimal daily scheduling of power generation in Hungary, and allow for a fast computerized solution of the problem.

A detailed description of the simplified model is presented in Chap. 4. The special forms of the objective function and the coefficient matrix of the linear constraints are presented.

Finally, in Chap. 5 a detailed numerical solution of the problem is presented. It is based on Benders' decomposition. Nonlinear constraints are linearized at some *working point*, and then the specially structured linear programming problem is solved by the aforementioned decomposition. Heuristics is used to find the next feasible working point.

The comprehensive appendix summarizes basic information about transmission networks in electric power systems. There are several approaches to mathematically describing transmission systems, and we have created our own version.

The book is largely based on the paper [12]. The references [10, 11] represents brief accounts of the main model, while [38] presents the first, albeit incomplete, model formulation.

In addition to the authors of this book, several other researchers participated in the project for shorter periods of time, providing us with help in designing algorithms, coding, and collecting data. We acknowledge the contributions of János Fülöp, Gerzson Kéri, László Sparing, Piroska Turchányi, and Béla Vizvári.

Piscataway, NJ, USA   András Prékopa
Zurich, Switzerland   János Mayer
Budapest, Hungary   Beáta Strazicky
Budapest, Hungary   István Deák
Budapest, Hungary   János Hoffer
Budapest, Hungary   Ágoston Németh
Budapest, Hungary   Béla Potecz
March 2014

Contents

Chapter 1
Control of Electric Power Systems

1.1 General Characteristics of Electric Power Systems

An electric power system is a combination of power-producing units, transmission lines, international cooperation, transformers, and a distribution network supplying customers with power under joint supervision and control.

The development of the system and its further modifications according to customers' long-term needs is the task of system planning. The task of *generation control* is, however, to operate the existing generating and electric apparatus as a whole at an optimal level.

This book is related to the subject of central generation control, which must meet the following main requirements:

1. The time-varying demand for active and reactive power must be met. The energy corresponding to active power cannot be stored in the system. This demand passes through the network in microseconds and manifests itself at the power plant generators. Fortunately, due to the composition of consumption, the power demand can be considered constant on the time scale of 1 min, and this is much larger than the reaction time of the controllers of the generators. The small fluctuations corresponding to time intervals shorter than 1 min are compensated by these devices. An additional characteristic of the system is that the production side has hardly any influence on consumption, which manifests itself for the production as an external factor.

2. Generated power should also comply with the quality requirements of a technical nature.

   - The frequency of the alternating current may only deviate from the nominal value (50 Hz) within a prescribed tolerance range. This is a significant factor with regard to the operation and stability of a system since in power plants power is generated by synchronous generators. In addition, some power consumption devices (synchronous motors, electric clocks, railway safety devices) are also calibrated for nominal frequency.

A. Prékopa et al., *Scheduling of Power Generation*, Springer Series in Operations Research and Financial Engineering, DOI 10.1007/978-3-319-07815-1_1,
© Springer International Publishing Switzerland 2014

- The voltages in a basic network must be within prescribed limits in order to have the voltage fluctuation within the tolerance range for power-consuming devices that are calibrated for a nominal voltage.
- The power supply must be continuous. An overload of a dangerous size should not occur, either in power generation or in the network apparatus. The daily configuration of the system must be devised such that any incidental breakdown can cause only minor damage.
- The international power exchange may differ from the scheduled values (determined by contracts) only within the tolerance range.
- Environmental pollution as a side effect should be minimized.

3. Production costs should be minimized.

Under these constraints, varying power demands can only be satisfied if the controllable quantities of the system, above all the amount of power injected into the network by power plants, are continually adjusted by operations control. Apart from this, operations control must determine in each of the periods the most appropriate values of the other variable system parameters, such as setting the position of tap-changing transformers, give orders to start up or shut down capacitors and shunt reactors, or change the network configuration. Concerning the devices mentioned, see Sect. 1.2.3 and Sect. A.2 of appendix.

Some of the foregoing questions are addressed in more detail in what follows.

1.1.1 Power Balance

The basic task of operations control is to satisfy the actual active and reactive power demand of consumers. Denoting the number of the nodes of the network by N (Sect. A.2 of appendix) and considering the networks up to the cutoff points of the international transmission lines, the following equations express the balance of generated and consumed power:

$$\sum_{i=1}^{N} P_{G_i} + \sum_{i=1}^{N} P_{T_i} = \sum_{i=1}^{N} P_{D_i} + P_v, \qquad (1.1.1)$$

$$\sum_{i=1}^{N} Q_{G_i} + \sum_{i=1}^{N} Q_{T_i} + \sum_{i=1}^{N} Q_{K_i} + \sum_{i=1}^{N} Q_{C_i} - \sum_{i=1}^{N} Q_{L_i} = \sum_{i=1}^{N} Q_{D_i} + Q_v. \qquad (1.1.2)$$

Equation (1.1.1) expresses the active power balance, whereas Eq. (1.1.2) expresses the reactive power balance. The applied notations are as follows (with index i indicating the serial numbers of the nodes):

P_{G_i}, Q_{G_i} = injection of active and reactive power by power plants;

P_{T_i}, Q_{T_i} = all active and reactive power arriving at connection points to international transmission lines;

P_{D_i}, Q_{D_i} = active and reactive power demand of consumers;

Q_{K_i} = reactive power generated by synchronous condensers;

Q_{C_i} = reactive power generated by capacitances of transmission lines and by shunt condensers;

Q_{L_i} = reactive power consumption by shunt reactors;

P_v, Q_v = overall loss of active respectively reactive power in transmission system.

The basic task of operations control is to choose the changeable parameters of the system in a such way that

(a) Eqs. (1.1.1) and (1.1.2) hold;
(b) The constraints listed at the beginning of Sect. 1.1 are satisfied;
(c) Parameter values are chosen from the set restricted by (a) and (b) that are the "most adequate" with respect to the whole system;
(d) The "most adequate" combination of changeable quantities and parameters is implemented at the right moment and the system is maintained in this state until the implementation of a new parameter setting becomes necessary.

The following questions have not yet received a definitive answer: how detailed should the mathematical model of an electric power system be and what criteria are needed to determine the so-called best combination of the regulated quantities and at the same time the best combination of system state characteristic quantities? Obviously, the direction of research will be greatly influenced by the experience gained with present electric power systems. For example, typically, greater breakdowns are followed by *critical investigations* of the existing operations control strategy.

An overall global optimization that takes into account the effect of all regulated quantities on all characteristic quantities of system states, especially if, beyond the usual constraints and economic efficiency aspects, it includes criteria concerning operation safety regulations, will constitute a very complex large-scale nonlinear problem. In principle, this problem may include statistical considerations, too. In the 1-day generation control problem, the influence of random effects can be disregarded in a first approximation, provided that the power demand is predicted with reasonable accuracy. In this book, an attempt is made to provide a deterministic formulation of the problem: a mixed-variable (discrete-continuous) model is built (Chap. 2).

1.1.2   Basic Elements of Present Generation Control Strategy

The present methods of generation control subdivide the task into more or less mutually independent separate subtasks, the latter themselves containing further simplifications.

The most common separation is based on the recognition that the active power balance, expressed in (1.1.1), is to a great extent independent of the reactive power and voltage conditions (Sect. A.4 of appendix). Therefore, the control of the active power and frequency can be separated from the control of the reactive power and voltage. The first one is often called $(P-f)$ *control*, whereas the latter is called $(Q-V)$ *control*. Their separability is highly supported by the practical aspect that opportunities for interference with power plants can also be decoupled accordingly: P_{G_i} production can be altered by modifying the basic signs of the turbine controllers (primary controllers), while Q_{G_i} injection can be changed by modifying the basic signs of the voltage controllers of generators.

Of the two types of control, $(P-f)$ control is the primary task because its economic effects are more obvious and because in modern power plants the tolerance range for the frequency is prescribed very strictly (in general, the tolerance in a normal mode of operation is $\pm 0.1\,\%$). To illustrate the mechanism of frequency control in a power plant, let us suppose that active power demand is decreasing somewhat at a certain time. Then the speed of rotation of the generator and the A.C. frequency increase to a certain extent as the torque generated by the turbine still corresponds to the earlier higher demand. The steam valve control mechanism of the turbine senses this increase and consequently decreases the amount of steam entering the turbine. In due course, the torque and the speed of rotation decrease, and the generator transmits less active power. As a result, a new dynamic equilibrium emerges.

The frequency control of the network is carried out in a way that (apart from those power plants in which the steam valve control mechanism of turbines with small reaction times are in operation) all the power plants inject active and reactive power in accordance with the schedule. Frequency control is performed by the selected power plants through their injection of active power.

In a normal mode of operation, system-level $(Q-V)$ control is generally given less attention. On the one hand, practically no production costs arise in connection with reactive power production respectively consumption; it plays only a minor economic role by slightly modifying transmission losses. On the other hand, when network voltages deviate from the nominal values, the tolerance level is at least $\pm 5\,\%$, and quite often a deviation of $\pm 10\,\%$ is still permitted. However, $(Q-V)$ control may play a significant role in restoring normal system operation during breakdowns.

In the sequel, generation control should always be understood as active power generation control.

1.1.3   Central Generation Control

Due to the complexity of power generation control outlined previously, the following subdivision based on time scale and tasks has been established for central generation control.

1.1.3.1   Subdivision by Time Scale

- Preparatory tasks:
  Calculations are done using estimated load data prior to actual generation and taking into account necessary maintenance work, that determine the basic conditions of online power generation, and voltage control (e.g., time points concerning startup and shutdown of individual blocks; daily generation and voltage schedule; export and import contracts; maintenance preparation).
- Generation control:
  Online (automatic or manual) control is performed by setting system power balance based on actual instantaneous data and having the following main constituents: on the one hand, automatic generation control (AGC) whose parts are the central control and the block control in power plants; on the other hand, automatic voltage control (AVC) or supervising voltage control (SVC) for setting the desired reactive power balance.

1.1.3.2   Subdivision by Task

- Ensuring power balance:
  P_{G_i} is determined in the AGC task in such a way that consumer demand is met in terms of the scheduled frequency and the exchange power complies with the value defined in the schedule.
- Decision making:
  In the AGC task the P_{G_i} gross load is subdivided among the available power plants, i.e., the active power to be generated by the individual power plants is determined. In the AVC (SVC) task the following factors must be determined: the power injection respectively voltage of the reactive power sources, the positions of the tap-changing transformers, and the start and stop states of individual network elements. This should be done in such a way that consumer demand is satisfied by the permitted bus voltages while at the same time the value of Q_{T_i} is kept within the acceptable range.
- Implementation:
  The changes in power production are imposed as determined in the AGC task. This is accomplished by the power plant controller and subsequently by the block controllers.

- In the AVC (SVC) task, orders are carried out concerning the injection of reactive power, modifications to the bus voltages, adjustment of transformer taps and performing start and stop actions. This is implemented by utilizing controllable sources of reactive power at the power stations, the excitation controls of the power plant generators and synchronous condensers, by the operation of the circuit breakers of the shunt reactors and condensers (high-voltage lines, cables), and by way of transformer tap adjustments.

1.2   Formulation of the Daily Scheduling Problem

Providing a daily schedule means the following things for each period of the day (periods of either 60 or 30 min):

- The most suitable mode of operation and the power level are determined for each power plant;
- The size of electric power export and import is determined; and
- The value of the potentials at the nodes are determined in such a way that the country should be supplied with electric energy, the voltages at the consumer nodes should deviate from the nominal value only slightly, transmission lines should not have a thermal overload, and power generation at the nodes connected to controllable sources of reactive power should fall within a preset range. Furthermore, practical specifications should be taken into account, including constraints concerning the power plants and the network.

In our case, schedules are for 25-h periods, i.e., 1 day and the first hour of the following day. Our model, which is presented in Chap. 2, is developed not merely to determine a sort of feasible daily schedule but to find a schedule in the set of feasible schedules that has the lowest possible cost with respect to a selected cost function.

In the following subsections the various elements of the preceding task are discussed. In the next subsection consumption and the daily demand curve are analyzed, followed by a subsection on power plants. The third subsection is devoted to a discussion of the transmission network, and in the last subsection the components of the cost of electric power production are examined.

1.2.1   Consumption and Daily Demand Curve

The main goal of a power supply is to provide consumers, who live in geographically distributed areas, with electric power. Apart from a few consumers (large industrial consumers), supply is provided by far-flung distribution networks having a relatively low voltage level that are connected by nodal substations to the basic transmission

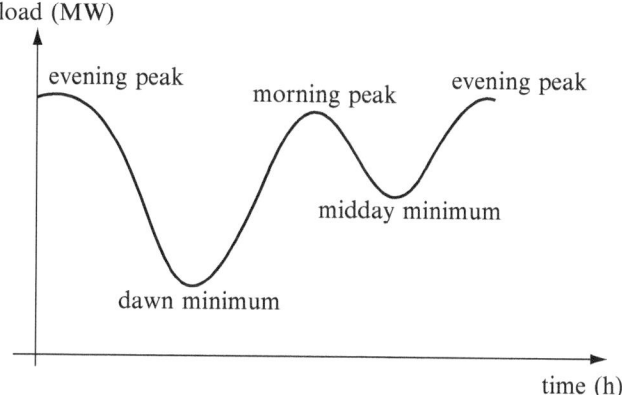

Fig. 1.1 Characteristic daily demand curve

network. The power demand at the nodes of the basic network varies significantly during the day. Within periods of 60 or 30 min the fluctuations of both the active and reactive power are fairly low; therefore, they will be considered as being constant.

The total amount of consumption in 1 h at the various nodes of the system, the amount of power consumed by the power plants themselves, the transmission losses, and the potential exported quantity together constitute the instantaneous power demand. The change in this power demand over time is described by the power demand as a function of time. In engineering terminology, this function is called the demand curve.

The daily demand curve depends in a complex way on the following factors:

- Seasons (e.g., time of sunrise and sunset, length of time between them, winter and summer time);
- Type of day (workday, holiday);
- Weather conditions (e.g., sunshine, temperature, clouds, wind);
- What's on TV.

In Hungary the daily power demand curve can be estimated fairly well, with an accuracy of 1–2 % using computer programs. This curve always includes two local minima and two local maxima.

Daily scheduling starts with the evening peak load period (from 5:00 to 7:00 p.m.) and involves 23 hourly periods and 4 periods of 30 min.

The power demand changes over the course of an hour only slightly, so in calculations it will be considered constant for the 60- and 30-min periods. This fact can also be expressed by stating that the demand curve is replaced by a piecewise constant function, also called a step function (Figs. 1.1 and 1.2).

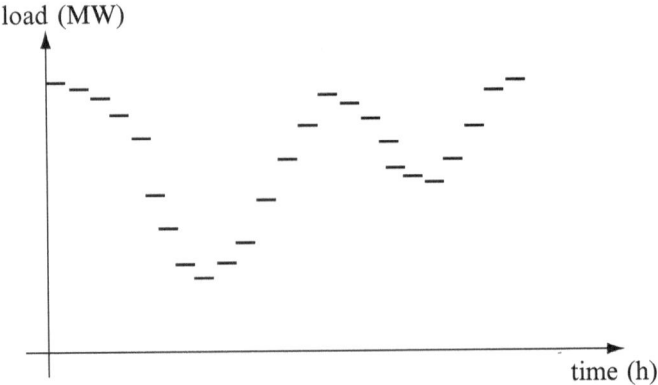

Fig. 1.2 Approximation of daily demand curve by a step function, the day being subdivided into hourly and half-hourly periods

1.2.2   Power Plants and Their Modes of Operation

On industrial scales, electric power is generated in thermal, hydroelectric, and nuclear power plants. In the various plants different economic and technical conditions prevail, depending on the applied modes of power conversion, the type and actual state of the machinery, the quality, and the accessibility of the primary energy source used.

In this book we consider electric energy production based solely on thermal power plants.

The different operating states of a power plant, corresponding to specific choices of the various types of equipment (groups of blocks, boiler–turbine–generator units), are called the *modes of operation* of a power plant. The modes of operation of a power plant operate in a predetermined range with respect to active and reactive power. The plant can switch from one mode of operation to another by virtue of *changeovers*. Changeovers are always carried out by starting/stopping some subdivisions of the apparatus. There are pairs of modes of operation that are based on entirely different equipment. Among these modes of operation, no changeover is permitted due to the very high costs this would entail.

For technical reasons, after shutting off a power plant unit, it can only return to the process of power generation following a standstill period of at least 4 h.

Import power is regarded as being produced by a separate power plant. The characteristics of the modes of operation of this so-called import power plant are in accordance with the conditions in interstate contracts.

The primary energy sources needed for the operation of power plants are brought to production sites as a result of cooperation among companies. The issues arising in connection with this cooperation (such as transport problems, storage in the case of hydroelectric power plants) may necessitate the restriction of the daily primary energy source consumption of a power plant or power plants. In mathematical terms this may mean either a lower or an upper bound. The consumption of primary energy sources can directly be expressed in terms of daily generated power; therefore, it may happen that lower or upper consumption bounds, unlike production bounds, are imposed on daily production. These are called *fuel constraints*.

1.2.3   Basic Electric Power System Network

Electric power generated in power plants is conveyed to consumers through an electric network. The branches of a transmission network can be classified in the following groups:

(a) Transmission lines of 750, 400, and 220 kV, connected to an international cooperation network and considered to extend as far as the border of the country;
(b) National basic network of main transmission lines of 400 and 220 kV;
(c) High-voltage main distribution network of 120 kV;
(d) Distribution network of a still lower voltage level.

The various parts of the network, corresponding to the different network voltage levels, are connected to one another by substations containing transformers. Depending on their power demand, consumers are connected to a network of the appropriate voltage level at the nodes.

For central generation control the electric power system manifests itself as a looped network also containing international connections, where consumption (demand) appears in concentrated form at high-voltage nodes, while power plants inject power into the network at high-voltage nodes that may partially be the same nodes where demand also appears. Accordingly, in a model for optimal daily scheduling, parts (a) and (b) of the preceding grouping of the transmission network are taken into account, together with those parts of the network under (c) (120 kV) to which significant power plants are connected or that form (or may form) shunt branches between nodes at 400- and 200-kV voltage levels. In the sequel, this part of the network will be considered the transmission network of the power system and will be called the basic network.

Power plants, reactive power sources, and consumers are connected to the nodes of the basic network. Power plants and consumption have already been discussed. Let us now give a brief account of the controllable devices that may inject or consume reactive power at the nodes (see also Sect. A.2 of appendix).

Devices acting as reactive power sources are connected to network nodes and can either inject (power plant generators, synchronous condensers, static or controllable capacitors) or consume (power plant generators, synchronous condensers, shunt reactors) reactive power. The power of those sources that can be controlled continuously can move freely within a specified range, while the power of those that can be switched (shunt reactors, static capacitors) changes proportionally to the square of the node voltage.

The nodes are connected to one another by branches that can be transmission lines, cables, or transformers. The branches are electrically symmetric with respect to their endpoints, and they have finite resistance, inductance, capacitance, and a finite thermal load bound. From the point of view of the generation control (and not for electrical engineering reasons) intersystem transmission lines are considered separately, connecting various national systems or systems of different Organizations. Due to the ohmic resistance of the branches, transmission is coupled with a loss depending on the load and on the voltage conditions. It should be emphasized that the elements of the transmission network are dynamic in the sense that, because of the fault of some network elements or their scheduled disconnection, the network configuration may change daily or possibly in even shorter intervals.

As an illustration, two daily periods with an extreme load are briefly discussed.

In a period of minimal night load (Sect. 1.2.1), transmission lines and cables are lightly loaded. Then their capacitive character is dominating, and as a result they generate reactive power. The overproduction of reactive power is accompanied by an increase in voltages (Sect. A.2 of appendix). The devices connected to the nodes that consume reactive power (see previous discussion) play an important role in compensating for this consumption.

The other extreme case is the peak load, when there is a high reactive power demand. On the one hand, this is because many power consumption devices (electric motors) consume reactive power. On the other hand, the transmission lines are highly loaded and their inductive character prevails. In this situation voltages drop, which can be compensated by reactive power sources.

Chapter 2
A General Mathematical Programming Model for the Scheduling of Electric Power Generation

In this chapter a general mathematical programming model of the scheduling problem as formulated in Sect. 1.2 is presented for the case where the electric power generation system includes thermal plants only. The model is called *general* because no simplifying assumptions are applied for the sake of obtaining a mathematical programming model tractable from the mathematical and computational viewpoints.

In Chap. 4 a simplified model will be presented. The two models differ a great deal with respect to the hypotheses concerning changeovers between different modes of operation.

In the general model, the impacts of changeovers between modes of operation can be summarized in the following way.

(a) Units that are shut off and units of the same kind that have not taken part in the operation previously can only enter the production process after a minimum 4 h standstill.

(b) Upon starting up a unit, startup costs arise depending on the length of the standstill period. No preliminary assumptions are made about the operational period of the units to be started. The length of the operational period is determined by the optimization over the whole day.

(c) Changeovers that may include the shutoff of certain devices and at the same time the startup of others are regarded as the sum of the two changeovers between the modes of operation. This rarely happens, but its possibility cannot be excluded.

Further differences between the two models include the following ones. While the general model can be formulated for a scheduling period of arbitrary length, the simplified model is designed for a 1-day (resp. 25-h) period because of the simplifying assumptions made in Sect. 3.2. The periods in the simplified model, where the power demand can be considered constant, have lengths of 30 or 60 min. In the general model, these periods are of arbitrary length. In the sequel, the term *period* will be used to mean a period of time for which the power demand is considered constant.

A. Prékopa et al., *Scheduling of Power Generation*, Springer Series in Operations Research and Financial Engineering, DOI 10.1007/978-3-319-07815-1__2,
© Springer International Publishing Switzerland 2014

2.1 Model Variables

In accordance with the problem formulation (Sect. 1.2), for each of the periods of the scheduling horizon we must determine the modes of operation to be used in the power plants, the power levels of the selected modes of operation, and the voltages at the nodes with controllable voltage.

Accordingly, the variables of the model will be the vectors denoting the modes of operation in the individual periods, the power levels, and the voltages. The large number of variables and the fact that we can refer to them in different ways present some difficulties in describing the model.

A general feature of our notation will be that the variable vectors and their components will be endowed by superscripts to denote the particular period to which the variable refers. If the superscript is missing, then a vector is being referred to, which arises by concatenating the corresponding vectors with superscripts based on the order of the periods. An exception to this convention will be the part devoted to the discussion of the voltage states of the network, where variable vectors without a superscript will denote variables corresponding to an arbitrary but fixed period. The discussion of the voltage conditions of the network is complicated enough, so to make it simpler, no superscript is included. When a variable without a superscript is used in this sense, it will be noted.

Let T denote in the sequel the number of periods of the scheduling time interval, while a_t denotes the length of the tth period. Accordingly, the length of the scheduling interval is $\sum_{t=1}^{T} a_t$. Let K denote the number of power plants and $M(k)$ the number of feasible modes of operation in the kth power plant.

For referring to the modes of operation of the power plants, the terminology first, second, ..., $M(k)$th mode of operation will be used in the case of the individual power plants. There is no restriction made with regard to the order of the modes of operation in the general model. Concerning this, see the notes at the end of Sect. 2.2.2.

Additional necessary notations will be defined at their first occurrence, and in Sect. 2.5 they will be summarized.

2.1.1 Mode-of-Operation Variables

The vector variables $\mathbf{y}^t$, $(t = 1, 2, \ldots, T)$, are introduced for the specification of the applicable modes of operation in power plants in different periods whose dimension is $\sum_{k=1}^{K} M(k)$. Furthermore, let $\mathbf{y}$ be a vector composed of the previous vectors via concatenation and having dimension $T \sum_{k=1}^{K} M(k)$.

The value of the components of $\mathbf{y}^t$ is 0 or 1. Their definition is as follows. To each of the power plants, and for a specific power plant to each of its modes of operation, there corresponds one of the components in the order of the power plants and for each power plant in the order of its feasible modes of operation. Based on this ordering, the first component of $\mathbf{y}^t$ corresponds to the jth mode of operation of the ith power plant for $j = l - \sum_{k=1}^{i-1} M(k)$, provided that $\sum_{k=1}^{i-1} M(k) < l \le \sum_{k=1}^{i} M(k)$ holds.

The components will often be referred to with a pair of indices, where the first index is the serial number of the power plant and the second one is the serial number of one of the modes of operation of that power plant corresponding to the particular component.

Let the value of the previously described first component be 1 if in the tth period in the ith power plant the jth mode of operation is to be used; otherwise let the value be 0.

Because exactly one mode of operation is active at one time, the preceding definition immediately implies that the following relations must hold:

$$\sum_{j=1}^{M(i)} y_{ij}^t = 1, \quad i = 1, 2, \ldots, K, \quad t = 1, 2, \ldots, T. \tag{2.1.1}$$

The vector $\mathbf{y}$ without a superscript is formed, according to our convention, by concatenating the $\mathbf{y}^t$, $t = 1, 2, \ldots, T$, vectors. Thus, for example, $y_l^{t_0} = y_{ij}^{t_0}$ corresponds to that component of the vector $\mathbf{y}$ that has the index

$$(t_0 - 1) \sum_{k=1}^{K} M(k) + l = (t_0 - 1) \sum_{k=1}^{K} M(k) + \sum_{k=1}^{i-1} M(k) + j.$$

Later on there will be a need for information on the modes of operation applied in the last period preceding the recent scheduling stage. This information can be provided by specifying the values of the variables of the modes of operation for this last period. Let $\mathbf{y}^0$ denote the corresponding vector. In the model this is a $\sum_{k=1}^{K} M(k)$-dimensional constant vector containing 0–1 components.

2.1.1.1 Remarks

1. The reader may wonder: what happens if in the power plants the set of feasible modes of operation is not the same in two consecutive scheduling stages, or if this set changes even within a single scheduling stage? (This might happen if, for example, some maintenance work is being finished at some time during the day and units not operable previously can now enter into production.)

   The answer is simple: in the scheduling stage the set of feasible modes of operation is specified in such a way that it should be the widest possible. Those modes of operation whose selection is only permitted in a couple of periods, are considered as separate modes of operation. Concerning these, we prescribe that in those periods where they are not available, the corresponding mode-of-operation variable take the value 0.

This convention makes it possible for the set of modes of operation $M(k)$, $k = 1, 2, \ldots, K$, to be independent of time and for each of the individual serial numbers of the modes of operation to represent the same mode of operation throughout.

2. The question may arise as to what happens if in a power plant just one of the modes of operation is feasible, that is, if $M(i) = 1$ holds for the ith power plant. In this case, in the model it is sufficient to specify the production level of this particular mode of operation; the corresponding mode-of-operation variable is superfluous since its value can only be 1.

   Despite this, to make the description of the model simpler, these power plants or modes of operation are not addressed separately. The variable y_{i1} is used instead, and the fulfillment of $y_{i1} = 1$ is ensured by applying Eq. (2.1.1).

3. Note that the mode-of-operation variables of the simplified model are defined in a different way.

2.1.2 Production-Level Variables

To specify the production levels of the modes of operation in the various periods, the production variable vectors $\mathbf{p}^t$ are used. Their dimension is $\sum_{k=1}^{K} M(k)$. Their concatenation ($t = 1, 2, \ldots, T$) results in vector $\mathbf{p}$, which has the dimension $T \sum_{k=1}^{K} M(k)$.

2.1.2.1 Definition of $\mathbf{p}^t$

To each mode of operation in each power plant we associate one of the components of $\mathbf{p}^t$ (in the order of the modes of operation and power plants exactly as was done in the case of the mode-of-operation variables). The components will also be referred to by double indices, where the first index is the serial number of the corresponding power plant and the second one is the serial number of the mode of operation. Accordingly, component p_{ij}^t is that component of vector $\mathbf{p}^t$ whose index is $\sum_{k=1}^{i-1} M(k) + j$ and whose index in vector $\mathbf{p}$ is $(t-1) \sum_{k=1}^{K} M(k) + \sum_{k=1}^{i-1} M(k) + j$.

For the permitted production levels minimal and maximal values are prescribed for each mode of operation at each power plant. These bounds will be denoted by $P_{ij}^{\min}$ and $P_{ij}^{\max}$, $i = 1, 2, \ldots, K$, $j = 1, 2, \ldots, M(i)$.

The value of the component p_{ij}^t is defined as follows. Let $p_{ij}^t = 0$ if the jth mode of operation is not applied at power plant i in period t. Otherwise, let its value be defined as the difference between the production level and its prescribed lower bound for the jth mode of operation at power plant i.

According to this definition, the production level of the jth mode of operation at power plant i can be specified as the sum

$$P_{ij}^{\min} y_{ij}^t + p_{ij}^t,$$

and the following relation must hold:

$$P_{ij}^{\min} y_{ij}^t \le P_{ij}^{\min} y_{ij}^t + p_{ij}^t \le P_{ij}^{\max} y_{ij}^t$$

respectively

$$0 \le p_{ij}^t \le (P_{ij}^{\max} - P_{ij}^{\min}) y_{ij}^t. \qquad (2.1.2)$$

According to the preceding definition, component p_{ij}^t can only have a nonzero value if $y_{ij}^t = 1$ holds. The fulfillment of this criterium is ensured by the factor y_{ij}^t in the product $P_{ij}^{\max} \cdot y_{ij}^t$ appearing on the right-hand side of the first inequality in (2.1.2). (Conversely, from the fulfillment of $y_{ij}^t = 1$ it does not follow that $p_{ij}^t > 0$ holds since if the mode of operation works on the permitted minimal level, then we have $p_{ij}^t = 0$.)

2.1.3 Voltage Variables

For all of the nodes of a transmission network, the real and imaginary parts of the complex voltage associated with the nodes are considered variables of the model. Let us denote in the tth period the real part of the voltages by $v_1^t, \ldots, v_N^t$ and the imaginary part by $w_1^t, \ldots, w_N^t$, where N stands for the number of the nodes of the network. Let the corresponding vectors be denoted by $\mathbf{v}^t$ and $\mathbf{w}^t$, respectively. In the sequel, in cases where a fixed period is considered, the superscripts will be omitted.

2.2 Objective Function of the Model

The objective function to be minimized is provided by the costs of electric power generation. This involves the cost of fuel needed to run the blocks of the power plants on a given level, the restart costs of the blocks consisting of heat losses due to stoppages and the deterioration costs due to changeovers, and, finally, the costs due to transmission losses. These three components are discussed separately.

2.2.1 Production Costs of the Power Plant Units

The costs of operation of power plant units can be specified in the following way.

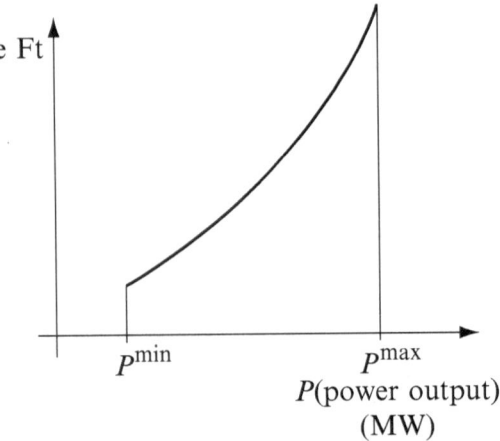

Fig. 2.1 Characteristic generation cost curve

To each feasible mode of operation of each power plant can be associated a characteristic cost curve (Fig. 2.1).

Let $f_{ij}(P)$ be a function defining the curve of costs, where P is the production level and the value of the function represents the costs of fuel consumption on production level P per unit time, and $f_{ij}(P)$ is a convex, strictly increasing function.

The domain of definition of $f_{ij}(P)$ is the interval $[P_{ij}^{\min}, P_{ij}^{\max}]$ because the mode of operation j in power plant i may only operate within these production limits. If another mode of operation is active, then the portion of production costs corresponding to mode of operation j at power plant i is obviously 0.

The following notations are introduced.

Let $K_{ij} = f_{ij}(P_{ij}^{\min})$, which represents the costs of minimal power output in the case where mode of operation j is active at power plant i.

Let $k_{ij}(P) = f_{ij}(P_{ij}^{\min} + P) - f_{ij}(P_{ij}^{\min})$, that is, the additional cost that arises when the amount P of electric power is generated beyond the minimal $P_{ij}^{\min}$ quantity if power plant i operates at production level $P_{ij}^{\min} + P$. The domain of the definition of function $k_{ij}(P)$ is the interval $[0, P_{ij}^{\max} - P_{ij}^{\min}]$.

According to our convention, the length of period t is denoted by a_t. Therefore, the production costs of mode of operation j at power plant i during period t can be formulated as

$$a_t \{ K_{ij} y_{ij}^t + k_{ij}(p_{ij}^t) \}. \tag{2.2.3}$$

Consequently, the partial costs, which have their origin in the operation of the power plant blocks, for the entire scheduling interval are

$$\sum_{t=1}^{T} a_t \sum_{i=1}^{K} \sum_{j=1}^{M(i)} \{ K_{ij} y_{ij}^t + k_{ij}(p_{ij}^t) \}. \tag{2.2.4}$$

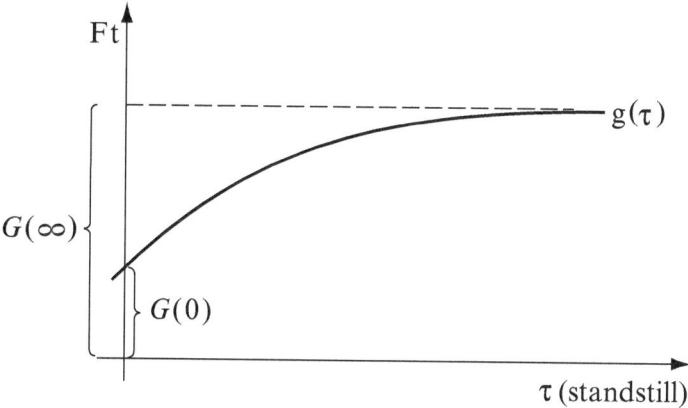

Fig. 2.2 Costs associated with stopping and restarting (standstill) of power plant units as a function of standstill time

2.2.2 Stoppage, Restart, and Changeover Costs

Considering the costs of stoppages of modes of operation means considering the costs that arise at the startup of power plant units following a standstill period. Regarding individual power plant units, these costs include deterioration costs and costs of extra maintenance work, as well as the costs of reheating following a standstill since during the standstill period the units cool down. Because heat loss has an exponential character, this type of cost can be expressed as an exponential function of the standstill time.

Figure 2.2 shows a cost curve representing the costs of restart of a unit following a standstill time τ. This cost curve can be specified by the formula

$$g(\tau) = G(0) + \{G(\infty) - G(0)\}(1 - e^{-c\tau}), \qquad (2.2.5)$$

where c, $G(0)$, and $G(\infty)$ are constant characteristic quantities of the power plant units; $G(0)$ represents deterioration and extra maintenance costs; $\{G(\infty) - G(0)\}(1 - e^{-c\tau})$ represents the costs due to heat loss; and $G(\infty)$ can be interpreted as the costs of a cold start following a long standstill. $c > 0$ holds since the costs of heat loss increase with the length of the standstill. The function $g(\tau)$ is defined for all $\tau \geq 0$ despite the fact that the units are either working continuously, thereby causing no stoppage or restart costs, or there is a standstill for at least 4 h. Due to the recommendations, patterns, guarantees, and operational specifications restricting transient heat stress, large-scale apparatus can only be restarted after a minimum standstill of 4 h.

In the model decisions are made concerning the operation respectively standstill of *modes of operation*. To take into account in the model the costs due to a standstill of *power plant units*, the connection between power plant units and modes of operation must be clarified. If every mode of operation meant the operation of different power plant units, then the standstill time of the mode of operation and the standstill time of the corresponding power plant units would be the same, implying that to compute the standstill costs of the power plant units, the standstill times of the modes of operation could be taken into account, and the latter could be specified as a function of the mode-of-operation variables y_{ij}^t. However, there is no such one-to-one correspondence between the modes of operation and the power plant units. The modes of operation are formed by technically feasible cooperations of the power plant units. Here is an example.

Suppose there are three units in a power plant, denoted by ①, ②, and ③. If Ⓐ and Ⓑ are two possible modes of operation in the power plant, where Ⓐ consists of the operation of unit ① and Ⓑ involves the cooperation of units ② and ③, the standstill time of the units is equal to the standstill time of the modes of operation, and the standstill costs are easily calculated. If, however, there is a third mode of operation, Ⓒ, involving the simultaneous operation of all three units, then the standstill time of the units is no longer equal to the standstill time of the modes of operation.

In the sequel we present a method for taking into account in the model the standstill costs of power plant units. For the description of this method the following terminology and notations will be used.

Regarding the jth mode of operation at the ith power plant, the starting time of the individual periods can be either *shutdown* time or *startup* time, while in the case of unchanged operations, the periods are called *continuous operation* periods respectively *unchanged inoperative* periods (or *continuous standstill* periods).

Therefore, the starting time of period t is a shutdown time if $y_{ij}^{t-1} = 1$ and $y_{ij}^t = 0$ hold. If $y_{ij}^{t-1} = 0$ and $y_{ij}^t = 1$, the starting time of the period is the startup time. Period t is a continuous operation period if $y_{ij}^{t-1} = 1$ and $y_{ij}^t = 1$ hold. Otherwise, if $y_{ij}^{t-1} = 0$ and $y_{ij}^t = 0$, then it is an unchanged inoperative period (since the vector $\mathbf{y}^0$ was defined earlier, the preceding terminology is well defined for $t = 1, 2, \ldots, T$).

Let us define the length of continuous standstills of the modes of operation in every period in the following way.

Let the length of a continuous standstill of the jth mode of operation at the ith power plant in period t be

- 0 if period t is a starting period or a continuous operation period;
- a_t if period t is a shutoff period;
- The length of time from the beginning of the last shutoff period preceding period t, including the length of period t, if period t is an unchanged inoperative period.

Let us denote by $\tau_{ij}(t)$ the continuous standstill time of mode of operation j at power plant i in period t. $\tau_{ij}(t)$ can be specified by the following formulas:

$\tau_{ij}(t) = \tau_{ij}(t-1) + a_t$ if period t is an unchanged inoperative period, i.e., if $y_{ij}^{t-1} = 0$ and $y_{ij}^t = 0$ hold;

$\tau_{ij}(t) = a_t$ if the starting time of period t is a shutdown time, i.e., if $y_{ij}^{t-1} = 1$ and $y_{ij}^t = 0$ hold;

$\tau_{ij}(t) = 0$ if period t is a continuous operation period or its starting time is a startup time, i.e., if either $y_{ij}^{t-1} = 1$ and $y_{ij}^t = 1$ or $y_{ij}^{t-1} = 0$ and $y_{ij}^t = 1$ hold.

The dependence of $\tau_{ij}(t)$ on the components of mode-of-operation variables can be expressed in the following product form as well:

$$\tau_{ij}(t) = \{\tau_{ij}(t-1) + a_t\}(1 - y_{ij}^t). \tag{2.2.6}$$

In fact, the value of this product is 0 if $y_{ij}^t = 1$ holds. Otherwise, $\tau_{ij}(t) = a_t$ if $y_{ij}^t = 0$ and $y_{ij}^{t-1} = 1$ since $\tau_{ij}(t-1) = 0$ holds. Finally, $\tau_{ij}(t) = \tau_{ij}(t-1) + a_t$ if $y_{ij}^t = 0$ and $y_{ij}^{t-1} = 0$ hold.

To complete the definition, the interpretation of $\tau_{ij}(0)$ must be supplied. Let $\tau_{ij}(0) = 0$ if $y_{ij}^0 = 1$ (i.e., if the mode of operation is active at the end of the previous planning stage), and let $\tau_{ij}(0)$ represent the length of the standstill preceding the planning stage if $y_{ij}^0 = 0$, i.e., $\tau_{ij}(0)$ equals $\tau_{ij}(T)$ of the previous period.

Like the modes of operation, the terms *shutdown* resp. *startup time, continuous operation* resp. *unchanged inoperative period* will be used with regard to the power plant units as well. The length of continuous standstills of power plant units will also be defined. To this end, the following notations will be used.

Let $N(i)$ be the number of units in power plant i. The units will be referred to by serial numbers; let $L(i) = \{1, 2, \ldots, N(i)\}$ be the index set of serial numbers. Let $J(i, j)$ denote that subset of $L(i)$ whose simultaneous operation of the units provides mode of operation j at power plant i.

A unit with serial number k_0 of the ith power plant is operative in period t if in the case where $y_{ij_0}^t = 1$ also $k_0 \in J(i, j_0)$ holds. If both $y_{ij_0}^t = 1$ and $k_0 \notin J(i, j_0)$ hold, then the k_0th unit is inoperative in period t.

The preceding information can also be specified with the value of the sum $\sum_{j:k_0 \in J(i,j)} y_{ij}^t$, where the addition is carried out for those values of j for which $k_0 \in J(i, j)$ holds:

$$\sum_{j:k_0 \in J(i,j)} y_{ij}^t = \begin{cases} 1 \text{ if unit } k_0 \text{ of power plant } i \text{ is operative in period } t, \\ 0 \text{ otherwise.} \end{cases}$$

The start time of period t is a shutdown time for unit k_0 at power plant i if

$$\sum_{j:k_0 \in J(i,j)} y_{ij}^{t-1} = 1 \quad \text{and} \quad \sum_{j:k_0 \in J(i,j)} y_{ij}^t = 0.$$

The terms *startup time, continuous operation* resp. *unchanged inoperative period* are defined in a similar way, i.e., regarding the sums $\sum_{j:k_0 \in J(i,j)} y_{ij}^t$ throughout.

We will need the lengths of continuous standstills of the power plant units. This is defined in an analogous way to the continuous standstill of the modes of operation.

Let $\xi_{ik}(t)$ denote the continuous standstill time of unit k at power plant i in period t. $\xi_{ik}(t)$ can be specified by the following formulas:

$$\xi_{ik}(t) = \xi_{ik}(t-1) + a_t \quad \text{if} \quad \sum_{j:k \in J(i,j)} y_{ij}^{t-1} = 0 \quad \text{and} \quad \sum_{j:k \in J(i,j)} y_{ij}^{t} = 0,$$

$$\xi_{ik}(t) = a_t \quad \text{if} \quad \sum_{j:k \in J(i,j)} y_{ij}^{t-1} = 1 \quad \text{and} \quad \sum_{j:k \in J(i,j)} y_{ij}^{t} = 0,$$

$$\xi_{ik}(t) = 0 \quad \text{if} \quad \sum_{j:k \in J(i,j)} y_{ij}^{t-1} = 1 \quad \text{and} \quad \sum_{j:k \in J(i,j)} y_{ij}^{t} = 1,$$

$$\text{or} \quad \sum_{j:k \in J(i,j)} y_{ij}^{t-1} = 0 \quad \text{and} \quad \sum_{j:k \in J(i,j)} y_{ij}^{t} = 1,$$

or it can also be expressed as the following product:

$$\xi_{ik}(t) = \{\xi_{ik}(t-1) + a_t\}\left(1 - \sum_{j:k \in J(i,j)} y_{ij}^{t}\right).$$

The value of $\xi_{ik}(0)$, which is needed to complete the definition, is determined in an analogous way to the definition of $\tau_{ij}(0)$.

It follows from the definition of the standstill times of the modes of operation and of units that

$$\xi_{ik}(t) = \min_{j:k \in J(i,j)} \tau_{i,j}(t), \qquad k = 1,2,\ldots,N(i), \quad i = 1,2,\ldots,K,$$

holds, and if the operation of a particular unit is necessary for the operation of only one mode of operation, then the standstill times of the unit and of the mode of operation are equal.

The costs of standstill in an electric power generation system are first considered periodwise, separately for the individual power plants resp. their units, and then these partial costs are summed up. Considering an individual power plant unit, for each period the costs arising in that particular period are taken into account.

Let $g_{ik}(\tau)$ denote the standstill-cost function of unit k at power plant i, $k = 1,2,\ldots,N(i), i = 1,2,\ldots,K$.

Let us examine the values of function $g_{ik}(\xi_{ik}(t))$. If the kth unit in the ith power plant is not operative in the tth period, then this function value is the cost of standstill corresponding to a standstill time till the end of that period. If the kth unit at the ith power plant is operative in the tth period, then $\tau_{ik}(t) = 0$ holds and the corresponding function value represents the deterioration costs.

Therefore, the costs due to the state of the kth unit of the ith power plant in the tth period can be specified utilizing the function $g_{ik}(\tau)$ as follows:

$$\left\{1-\left(\sum_{j:k\in J(i,j)} y_{ij}^{t-1}\right)\left(\sum_{j:k\in J(i,j)} y_{ij}^{t}\right)\right\} g_{ik}(\xi_{ik}(t))-\left\{1-\sum_{j:k\in J(i,j)} y_{ij}^{t}\right\} g_{ik}(\xi_{ik}(t-1)).$$

$$(2.2.7)$$

To see this, let us discuss the following four cases.

(a) If the kth unit of the ith power plant is inoperative in the tth period, as well as in the preceding period, i.e., if

$$\sum_{j:k\in J(i,j)} y_{ij}^{t-1} = 0 \quad \text{and} \quad \sum_{j:k\in J(i,j)} y_{ij}^{t} = 0$$

hold, then the value of expression (2.2.7) is

$$g_{ik}(\xi_{ik}(t)) - g_{ik}(\xi_{ik}(t-1)).$$

This is the cost of being in an unchanged inoperative state in a time interval of length a_t, from time $\xi_{ik}(t-1)$ to time $\xi_{ik}(t)$, in period t.

(b) If for the kth unit in the ith power plant the starting point of the tth period is a shutdown time, that is, in the $(t-1)$th period a mode of operation is active that requires the operation of the kth unit, $(\sum_{j:k\in J(i,j)} y_{ij}^{t-1} = 1)$, and in the tth period none of this type of mode of operation is active, $(\sum_{j:k\in J(i,j)} y_{ij}^{t} = 0)$, then the value of expression (2.2.7) is

$$g_{ik}(a_t) - g_{ik}(0),$$

which represents the costs of standstill for a time interval of length a_t.

(c) If the kth unit at the ith power plant is operative in the tth period as well as in the $(t-1)$th period, $(\sum_{j:k\in J(i,j)} y_{ij}^{t-1} = 1$ and $\sum_{j:k\in J(i,j)} y_{ij}^{t} = 1)$, then the value of (2.2.7) is 0.

(d) If for the kth unit at the ith power plant the starting point of the tth period is a startup time, that is, the unit is active in the tth period, $(\sum_{j:k\in J(i,j)} y_{ij}^{t} = 1)$, but it is not in operation in the preceding period, $(\sum_{j:k\in J(i,j)} y_{ij}^{t-1} = 0)$, then the value of (2.2.7) is $g_{ik}(0)$, representing the deterioration costs corresponding to the preceding standstill time interval.

Finally, by summing up the values in (2.2.7) for $t = 1, 2, \ldots, T$, we arrive at the overall cost for the entire scheduling interval, and this corresponds to the states of the kth unit at the ith power plant. This is represented as a sum of function-value increments, corresponding to argument increments having lengths equal to the period lengths. The deterioration cost $g_{ik}(0)$ is taken into account at the time of startup of the unit (Fig. 2.3). In continuous operation periods, no standstill costs arise.

Let us give an illustration with a specific example. Assume there are three power plant units at the fifth power plant, where the simultaneous operation of units 1 and 3

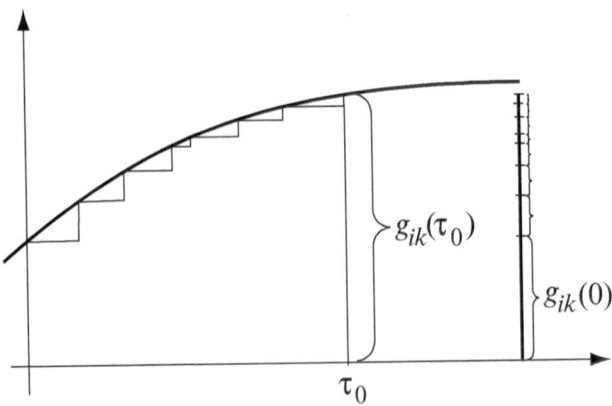

Fig. 2.3 Representation of cost of changeover corresponding to standstill τ_0 as a sum of function-value increments and $g_{ik}(0)$

provides mode of operation 1, and the operation of all three units provides mode of operation 2: $M(5) = 2$, $N(5) = 3$, $J(5,1) = \{1,3\}$, $J(5,2) = \{1,2,3\}$.

Let us further assume that mode of operation 1 is active in the period preceding the current planning stage and in the first three periods of the planning stage as well, and in the next four periods mode of operation 2 is active. Finally, mode of operation 1 is operative till the end of the planning stage. Let the planning stage contain ten periods. Let us suppose that mode of operation 2 was shut off 2 h prior to the planning stage. Then the values of the mode-of-operation variables are as follows:

$$y_{51}^0 = 1,\ y_{51}^1 = 1,\ y_{51}^2 = 1,\ y_{51}^3 = 1,\ y_{51}^4 = 0,\ y_{51}^5 = 0,\ y_{51}^6 = 0,\ y_{51}^7 = 0,$$
$$y_{51}^8 = 1,\ y_{51}^9 = 1,\ y_{51}^{10} = 1;$$
$$y_{52}^0 = 0,\ y_{52}^1 = 0,\ y_{52}^2 = 0,\ y_{52}^3 = 0,\ y_{52}^4 = 1,\ y_{52}^5 = 1,\ y_{52}^6 = 1,\ y_{52}^7 = 1,$$
$$y_{52}^8 = 0,\ y_{52}^9 = 0,\ y_{52}^{10} = 0.$$

The lengths of the standstill intervals of the modes of operation, $\tau_{51}(t)$ and $\tau_{52}(t)$, $t = 0,1,2,\ldots,10$, are as follows:

$$\tau_{51}(0) = 0, \qquad \tau_{51}(1) = \tau_{51}(2) = \tau_{51}(3) = 0,\ \tau_{51}(4) = a_4,\ \tau_{51}(5) = a_4 + a_5,$$
$$\tau_{51}(6) = a_4 + a_5 + a_6,\ \tau_{51}(7) = a_4 + a_5 + a_6 + a_7,\ \tau_{51}(8) = 0,\quad \tau_{51}(9) = 0,$$
$$\tau_{51}(10) = 0,$$

$$\tau_{52}(0) = 2, \ \tau_{52}(1) = 2 + a_1, \ \tau_{52}(2) = 2 + a_1 + a_2, \ \tau_{52}(3) = 2 + a_1 + a_2 + a_3,$$

$$\tau_{52}(4) = \tau_{52}(5) = \tau_{52}(6) = \tau_{52}(7) = 0, \ \tau_{52}(8) = a_8, \ \tau_{52}(9) = a_8 + a_9,$$

$$\tau_{52}(10) = a_8 + a_9 + a_{10}.$$

The lengths of the standstill intervals of the power plant units, $\xi_{51}(t)$, $\xi_{52}(t)$, and $\xi_{53}(t)$, $t = 0,1,2,\ldots,10$, are as follows:

$$\xi_{51}(t) = \min_{j:1 \in J(i,j)} \tau_{5,j}(t) = \begin{cases} \tau_{51}(t) = 0, \ t = 0,1,2,3,8,9,10, \\ \tau_{52}(t) = 0, \ t = 4,5,6,7; \end{cases}$$

$$\xi_{52}(t) = \min_{j:2 \in J(5,j)} \tau_{5,j}(t) = \tau_{52}(t);$$

$$\xi_{53}(t) = \min_{j:3 \in J(5,j)} \tau_{5,j}(t) = \min(\tau_{51}(t), \tau_{52}(t)) = 0.$$

Therefore, the costs due to the periodwise states of the power plant units are as follows:

Period	Unit 1	Unit 2	Unit 3
1.	0	$g_{52}(2 + a_1) - g_{52}(2)$	0
2.	0	$g_{52}(2 + a_1 + a_2) - g_{52}(2 + a_1)$	0
3.	0	$g_{52}(2 + a_1 + a_2 + a_3) - g_{52}(2 + a_1 + a_2)$	0
4.	0	$g_{52}(0)$	0
5.	0	0	0
6.	0	0	0
7.	0	0	0
8.	0	$g_{52}(a_8) - g_{52}(0)$	0
9.	0	$g_{52}(a_8 + a_9) - g_{52}(a_8)$	0
10.	0	$g_{52}(a_8 + a_9 + a_{10}) - g_{52}(a_8 + a_9)$	0

The sum of the standstill costs for all of the periods is

$$g_{52}(2 + a_1 + a_2 + a_3) - g_{52}(2) + g_{52}(0) + g_{52}(a_8 + a_9 + a_{10}) - g_{52}(0).$$

This is in agreement with the fact that power plant unit 2 is in standstill for 2 h in the preceding planning stage and in the first three periods of the recent planning stage $[g_{52}(2 + a_1 + a_2 + a_3) - g_{52}(2)]$, then it is started $[g_{52}(0)]$ and is again in standstill from the eighth up to the tenth period $[g_{52}(a_8 + a_9 + a_{10}) - g_{52}(0)]$.

The total cost of standstill in the entire electric power system is

$$\sum_{t=1}^{T} \sum_{i=1}^{K} \sum_{k=1}^{N(i)} \left[\left\{ 1 - \left(\sum_{j:k \in (i,j)} y_{ij}^{t-1} \right) \left(\sum_{j:k \in J(i,j)} y_{ij}^{t} \right) \right\} g_{ik}(\xi_{ik}(t)) \right]$$

$$-\left\{1 - \sum_{j:k\in J(i,j)} y_{ij}^t\right\} g_{ik}(\xi_{ik}(t-1))\Bigg]. \tag{2.2.8}$$

This sum does not contain the deterioration costs of units due to standstill in the last period of the considered planning stage. This portion of costs is taken into account at the time of restart, but now no restart occurs. The deterioration costs of the units in standstill in the last period are

$$\sum_{i=1}^{K}\sum_{k=1}^{N(i)} g_{ik}(0)\left(1 - \sum_{j:k\in J(i,j)} y_{ij}^T\right), \tag{2.2.9}$$

which is taken into account in the subsequent planning stage.

In calculating the standstill costs, is not taken into account that the actual length of standstill of the stopped equipment is always at least 4 h. The fulfillment of this requirement will be ensured in the specification of the model constraints.

2.2.2.1   Remark

The way we have considered the costs of standstill and restart is more general than is needed for the operations management of electric power plants. It is certainly true that the units of a specific power plant may cooperate in many different ways, that is, several different modes of operation are feasible at power plants, and those modes can be specified as index sets consisting of the serial numbers of the corresponding power plant units (these have been denoted by $J(i,j)$, $j = 1,2,\ldots,M(i)$). However, according to operations management practice, for a given planning cycle only a subset of all possible modes of operation is feasible, namely a subset that can be ordered in such a way that for the corresponding $J(i,j)$ sets

$$J(i,j_1) \supset J(i,j_2) \supset \cdots \supset J(i,j_r)$$

holds. Sometimes (rarely, in fact), a more general case occurs when the acceptable modes of operation in the planning stage can only be ordered in such a way that feasible $J(i,j)$ sets can be subdivided into two groups of the aforementioned character. Let these groups be

$$J(i,j_1) \supset J(i,j_2) \supset \cdots \supset J(i,j_r),$$
$$J(i,l_1) \supset J(i,l_2) \supset \cdots \supset J(i,l_s).$$

In general, these two groups are not independent. $J(i,l_1)$ is part of a set of indices of the first group, and some set of the first group is a subset of $J(i,l_s)$. Based on the technical and technological conditions, this is the most general case that can actually occur in practice.

We provide examples for both cases:

Let us suppose that there are five units in a power plant and the $J(i,j)$ sets corresponding to the modes of operation provided by the possible forms of cooperation of these units are as follows:

$$\{1\}, \{1,2\}, \{1,3\}, \{1,2,4\}, \{1,3,5\}, \{1,2,3,5\}, \{1,2,3,4,5\}.$$

An example of the first case is a planning stage where the following modes of operation are possible:

$$\{1\}, \{1,2\}, \{1,2,3,5\}, \{1,2,3,4,5\}$$

as it is true that

$$\{1,2,3,4,5\} \supset \{1,2,3,5\} \supset \{1,2\} \supset \{1\}.$$

An example of the second case is a planning stage where the following modes of operation are possible:

$$\{1\}, \{1,2\}, \{1,3\}, \{1,3,5\}, \{1,2,3,5\} \text{ and } \{1,2,3,4,5\}.$$

Then

$$\{1,2,3,4,5\} \supset \{1,2,3,5\} \supset \{1,2\} \supset \{1\}; \quad \{1,3,5\} \supset \{1,3\}$$

and

$$\{1,2,3,4,5\} \supset \{1,3,5\}; \quad \{1,3\} \supset \{1\}$$

are satisfied.

A planning stage where, in addition to the modes of operation in the foregoing second example, the mode of operation $\{1,2,4\}$ is also feasible cannot occur for engineering reasons. However, there can be another planning stage where the modes of operation $\{1\}, \{1,2\}, \{1,2,4\}$ are feasible. This would again be an example of the first case.

Since in this book a model for a given scheduling period is considered, $M(i)$, $i = 1,2,\ldots,k$, indicates the number of modes of operation that are *feasible in the given scheduling period* and not the number of all possible modes of operation. Thus, in the calculation of the costs of standstill, the previously given characteristics of the sets $J(i,j)$, $j = 1,2,\ldots,M(i), i = 1,2,\ldots,K$, can be used.

In the simplified model, it will be assumed that the order of the modes of operation is as follows:

$$J(i,1) \supset J(i,2) \supset \cdots \supset J(i,M(i)).$$

2.2.3 Costs from Transmission Losses

The electric energy generated by power plants is conveyed to consumers through an electric network system. Due to the ohmic resistance of the branches of the network, transmission is accompanied by a loss of active power. Part of the electric energy is converted into heat. Taking a fixed period, costs that originate in transmission losses are

$$C^v(\mathbf{v}, \mathbf{w}) = \gamma a_t P^v(\mathbf{v}, \mathbf{w}), \qquad (2.2.10)$$

where $C^v(\mathbf{v}, \mathbf{w})$ is the value of transmission losses in the local currency (Fts) for the given period; γ is the cost of 1 MWh in the local currency (Fts); a_t is the length of period t in hours; $P^v(\mathbf{v}, \mathbf{w})$ is in turn the total active power loss in the network in megawatts. The function $P^v(\mathbf{v}, \mathbf{w})$ is a convex quadratic function of the variables $\mathbf{v}, \mathbf{w}$ [see (A.4.21), (A.4.30), and (A.4.31)]. The total cost of transmission losses is the sum of the costs of losses over the periods.

2.3 Model Constraints

The system of model constraints contains, on the one hand, constraints that are necessary for the definition of the variables and, on the other hand, constraints that contribute to the description of the operation of the electric power system. Constraints can also be grouped on the basis of whether they represent relationships among the variables of an individual period in a repeated fashion for the periods or whether they prescribe relationships among the variables of several periods.

2.3.1 Constraints Repeated for Periods

For a given period we require that only one mode of operation may be active at each of the power plants, the power level of the mode of operation should be within the permitted range, and the amount of generated electric power should be equal to the nationwide demand for electric power, plus the transmission losses and self-consumption needed for electric power generation. We also require that the network conditions of the electric power system must be feasible in every period.

As a result of the definition of the mode-of-operation variables, the requirement of *one mode of operation in one power plant* means that Eqs. (2.1.1) should hold. These conditions will be referred to as the *special ordered set* (SOS) *constraints*, which is a generally accepted term in discrete programming. A SOS is a set of variables having a value of 0 or 1 such that among the variables one and only one has a value of 1.

Naturally, because of the definition of the mode-of-operation variables, the

$$y_{ij}^t = 0 \quad \text{or} \quad 1, \quad j = 1, 2, \ldots, M(i), \quad i = 1, 2, \ldots, K$$

criteria must be met for these variables, too.

In accordance with the definition of production-level variables, inequalities (2.1.2) must be satisfied in every period. In the sequel, these inequalities will be referred to as *variable coupling constraints*. These are constraints that directly *couple* the mode-of-operation and production-level variables for each power plant and for each plant's mode of operation.

The constraints associated with the requirement of *satisfying the power demand* are called *supply constraints* (to every period is attached one constraint of this sort). With the help of production-level and mode-of-operation variables, the production level of the electric power system in period t can be specified by the following sum:

$$\sum_{i=1}^{K} \sum_{j=1}^{M(i)} (P_{ij}^{\min} y_{ij}^t + p_{ij}^t). \tag{2.3.11}$$

A so-called function of self-consumption is provided for every mode of operation at every power plant. It determines the amount of electric power necessary for the operation of the power plant taken as a function of the power level of the mode of operation. Let P_{ij}^{self} denote the function of self-consumption belonging to the jth mode of operation at power plant i. The self-consumption of power plant i in period t is

$$\sum_{j=1}^{M(i)} P_{ij}^{\text{self}} (P_{ij}^{\min} y_{ij}^t + p_{ij}^t), \tag{2.3.12}$$

and the self-consumption of the entire electric power system in period t is

$$P^{t \text{ self}} = \sum_{i=1}^{K} \sum_{j=1}^{M(i)} P_{ij}^{\text{self}} (P_{ij}^{\min} y_{ij}^t + p_{ij}^t). \tag{2.3.13}$$

Let $P^{t \text{ dem}}$ denote the value of electric power demand in period t and $P^{t \text{ loss}}$ the transmission losses in period t.

The *supply constraint* is as follows:

$$\sum_{i=1}^{K} \sum_{j=1}^{M(i)} \left(P_{ij}^{\min} y_{ij}^t + p_{ij}^t \right) = P^{t \text{ dem}} + P^{t \text{ loss}} + P^{t \text{ self}}. \tag{2.3.14}$$

Note that $P^{t\,\text{loss}}$ and $P^{t\,\text{self}}$ are actually functions, with the values of $P^{t\,\text{loss}}$ being specified by a convex quadratic function of the voltage variables $\mathbf{v}^t$ and $\mathbf{w}^t$ and the values of $P^{t\,\text{self}}$ depending nonlinearly on the $\mathbf{p}^t$ and $\mathbf{y}^t$ variables. In contrast, $P^{t\,\text{dem}}$ is a constant provided by the demand curve. Methods for calculating loss and self-consumption are not detailed here. In Chap. 3 we will make some simplifying assumptions concerning them as well.

Constraints describing the network conditions of the electric power system will be discussed next (where the superscripts identifying the periods will be omitted).

To build up a system of constraints related to the transmission network, as a starting point we consider relations (A.3.43), which establish a connection between the power injections at the nodes and the voltages of the nodes. For ease of discussion we reproduce them here:

$$\left.\begin{array}{l} f_i(\mathbf{v},\mathbf{w}) = P_i \\ g_i(\mathbf{v},\mathbf{w}) = Q_i \end{array}\right\} \quad i = 1,\dots,N. \tag{2.3.15}$$

In the sequel, let I_E denote the set of serial numbers of those nodes to which power plants are connected, and let N_E be the number of elements of I_E. Let us denote further by I_M the set of serial numbers of those nodes to which a piece of equipment of the electric power system is connected for which the value of the reactive power injection respectively consumption is not prescribed for the given period; it can vary within a given range. Later on these nodes will be called nodes connected to controllable sources of reactive power, or reactive power source nodes for short. (For more details see Sect. A.3 of appendix). Let N_M denote the number of elements of I_M. The index sets I_E and I_M are considered to be ordered according to the increasing order of the serial numbers of their elements.

In what follows, the P_i active power appearing in Eqs. (2.3.15) will be considered in the decomposed form

$$P_i = P_i^G - P_i^F, \quad i = 1,\dots,N, \tag{2.3.16}$$

where superscript G represents generated (injected) power and F indicates consumption. The quantities P_i^F, $i = 1,\dots,N$, are the values of consumption (power demand) at the nodes, and obviously, the relation $P_i^G = 0$, $i \notin I_E$, holds.

Let $\hat{\mathbf{P}}^G$ denote the N_E-dimensional vector formed by the $i \in I_E$ components of the vector $\mathbf{P}^G$. To include Eqs. (2.3.15) in the model as constraints, it is sufficient to express the vector $\hat{\mathbf{P}}^G$ in terms of the vectors $\mathbf{p}$ and $\mathbf{y}$, the latter denoting vectors composed of production-level respectively mode-of-operation variables. The dimension of $\mathbf{p}$ (and $\mathbf{y}$ as well) will be denoted by n.

The amount of power actually generated in the different modes of operation (denited by $\hat{\mathbf{p}} \in R^n$) can be obtained as

$$\hat{\mathbf{p}} = \mathbf{p} + \text{diag}\,(\mathbf{p}^{\text{min}})\mathbf{y}, \tag{2.3.17}$$

where diag $(\mathbf{p}^{\min})$ is a diagonal matrix whose diagonal elements are the components of $\mathbf{p}^{\min}$, where $\mathbf{p}^{\min}$ is the vector of the minimal power levels corresponding to power plants and their modes of operation.

Consequently, active power injection at individual nodes can simply be obtained by adding up the actually produced powers corresponding to the modes of operation of those power plants that are connected to that node, in other words, by adding up the corresponding components of $\hat{\mathbf{p}}$. This means carrying out the following linear transformation:

$$\hat{\mathbf{P}}^{G} = \mathbf{H}\hat{\mathbf{p}}, \tag{2.3.18}$$

where $\mathbf{H}$ is a summation matrix of size $(N_E \times n)$. In the ith row of the matrix is a 1 in positions corresponding to the modes of operation of the power plant connected to the ith power plant node; all the other elements in the row are 0; $i = 1, \ldots, N_E$.

Let $\mathbf{D}$ be the matrix of the linear transformation defined by (2.3.17) and (2.3.18). Its size is $N_E \times (2n)$. Furthermore, let $\mathbf{D}^{T} = (\mathbf{d}_1, \ldots \mathbf{d}_L)$. Thus, we have the following relation:

$$\hat{\mathbf{P}}^{G} = \mathbf{D}\begin{pmatrix} \mathbf{p} \\ \mathbf{y} \end{pmatrix}. \tag{2.3.19}$$

If $i \in I_E$, let $E(i)$ denote the serial number of index i within the ordered set I_E.

Before the specification of the final form of the system of constraints (2.3.15) to be attached to the model, one more comment is necessary. In Eqs. (2.3.15), in the case of $i \in I_M$, the quantities Q_i are to be treated as variables (Sect. A.3 of appendix). However, because they occur only on the right-hand side of the reactive-power relations, constraints with serial numbers $i \in I_M$ can be omitted from those in the reactive power part.

Therefore, the system of constraints connecting the $\mathbf{p}$, $\mathbf{y}$ variables to the voltages at the nodes takes the form

$$\mathbf{d}_{E(i)}^{T}\begin{pmatrix} \mathbf{p} \\ \mathbf{y} \end{pmatrix} - f_i(\mathbf{v}, \mathbf{w}) = P_i^{F}, \qquad i \in I_E,$$

$$-f_i(\mathbf{v}, \mathbf{w}) = P_i^{F}, \qquad i \notin I_E, \tag{2.3.20}$$

$$g_i(\mathbf{v}, \mathbf{w}) = Q_i, \qquad i \notin I_M,$$

where the functions $f_i(\mathbf{v}, \mathbf{w})$, $g_i(\mathbf{v}, \mathbf{w})$ are quadratic functions of the variables $\mathbf{v}$, $\mathbf{w}$ [see (A.3.42)], and constant quantities appear on the right-hand side of the constraints.

The rest of the system of constraints with respect to the network express restrictions in the form of inequalities concerning the various electric quantities. Each one is discussed separately.

2.3.1.1   Voltage Limit Constraints

In an electric power system every power consumption device is calibrated for a fixed nominal voltage level. A too large deviation from this level may result in damage or equipment malfunction. To keep consumers' voltage near the nominal value, the voltage at the nodes of a high-voltage network should be within given intervals. Let $V_i^{\min}$ be the lower bound on the absolute value of voltage at the ith node of a high-voltage transmission network, and let the upper bound be $V_i^{\max}$, $i = 1, \ldots, N$. Then the voltage limit constraints are as follows:

$$V_i^{\min} \leq (v_i^2 + w_i^2)^{1/2} \leq V_i^{\max}, \qquad i = 1, \ldots, N. \tag{2.3.21}$$

2.3.1.2   Branch-Load Constraints

Branch-load constraints refer to the thermal loadability of branches and prevent overheating. For each of the branches (i, k) of the transmission network a $T_{ik}^{\max}$ value is specified. For safety reasons, the absolute value of power (apparent power) flowing in the branch is not allowed to exceed this value. Power flow is discussed in Sect. A.4 of appendix; the apparent power is connected to the model variables by Eq. (A.4.20). In this respect the power loss in the branch can be neglected; consequently, the branch load constraints are as follows:

$$|S_{ik}(v_i, v_k, w_i, w_k)| \leq T_{ik}^{\max} \tag{2.3.22}$$

for every (i, k) branch of the transmission network.

2.3.1.3   Reactive Power Injection Constraints

These constraints restrict the injection or consumption of reactive power at the nodes connected to controllable sources of reactive power (in our case, their serial numbers are $i \in I_M$). In the case of power plants and synchronous condensers, depending on the characteristics of the synchronous generators, their reactive power injection (resp. consumption) is limited and implies limits concerning the reactive power injection respectively consumption at the corresponding node. There are lower and upper bounds specified for the other controllable sources of reactive power regarding their injection respectively consumption of reactive power (concerning the controllable sources of reactive power see Sect. A.2 of appendix).

If the ith node is a controllable source of reactive power and the bounds on reactive power at the power plant are $Q_i^{\min}(\mathbf{y})$, $Q_i^{\max}(\mathbf{y})$, $i \in I_M \cap I_E$, otherwise $Q_i^{\min}$, $Q_i^{\max}$, $i \in I_M \setminus I_E$, then the following constraints arise:

$$Q_i^{\min}(\mathbf{y}) \leq Q_i \leq Q_i^{\max}(\mathbf{y}), \qquad i \in I_M \cap I_E, \tag{2.3.23}$$

$$Q_i^{\min} \le Q_i \le Q_i^{\max}, \qquad i \in I_M \setminus I_E.$$

As stated earlier, the quantities Q_i are specified in explicit form in Eqs. (2.3.15); therefore, the reactive power injection constraints, imposed via substitutions, are as follows:

$$Q_i^{\min}(\mathbf{y}) \le g_i(\mathbf{v}, \mathbf{w}) \le Q_i^{\max}(\mathbf{y}), \qquad i \in I_M \cap I_E,$$

$$Q_i^{\min} \le g_i(\mathbf{v}, \mathbf{w}) \le Q_i^{\max}, \qquad i \in I_M \setminus I_E. \tag{2.3.24}$$

In the inequalities (2.3.24), in the lines referring to the power plants, the bounds on both the left- and right-hand sides are functions of the mode-of-operation variables since the bounds related to reactive power injection can be determined by adding up the bounds related to the reactive power injection of the modes of operation of the power plants that are connected to the nodes. Let $\hat{\mathbf{Q}}_i^{\min}$ and $\hat{\mathbf{Q}}_i^{\max}$ represent the following vectors: their dimension equals the dimension of $\mathbf{y}$, and in the position of the modes of operation that are connected to the ith node there are lower and upper bounds attached to the corresponding mode of operation. The remaining components are 0 in both vectors. Therefore, the realization of the aforementioned addition leads to the following bounds on the controllable sources of reactive power:

$$Q_i^{\min}(\mathbf{y}) = \mathbf{y}^T \hat{\mathbf{Q}}_i^{\min}, \quad Q_i^{\max}(\mathbf{y}) = \mathbf{y}^T \hat{\mathbf{Q}}_i^{\max}, \quad i \in I_M \cap I_E.$$

2.3.2 Constraints Connecting the Periods

Among the variables of the different periods of the model, the *stop-and-start constraints* and the *fuel constraints* establish links.

In the model, the stop-and-start constraints involve mode-of-operation variables only, and each fuel constraint contains the production-level and mode-of-operation variables of just one power plant. Constraints involving voltage variables of several periods are not included in the model.

In the *stop-and-start constraints*, for technical reasons, shutoff equipment cannot become operative before a minimum standstill of 4 h, and for economic reasons a power plant unit having the same characteristics as the shutoff device cannot start either. Technically, this latter case would be possible, but for economic reasons it is not sensible to stop a unit if, for example, a unit with similar characteristics and production level will have to be started 2 h later.

The requirement ensuring a minimum standstill of the shutoff units of 4 h can be formulated using the $\xi_{ik}(t)$ standstill functions defined in Sect. 2.2.2 by requiring that the value of the product

$$\xi_{ik}(t-1)\left(1 - \sum_{j:k\in J(i,j)} y_{ij}^{t-1}\right)\left(\sum_{j:k\in J(i,j)} y_{ij}^{t}\right) \qquad (2.3.25)$$

be either 0 or otherwise at least 4. This must hold for $t = 1, 2, \ldots, T$, $i = 1, 2, \ldots, K$, and $k = 1, 2, \ldots, N(i)$. Such a formulation of the condition is advantageous because the units shut off at the end of the preceding planning stage and, prior to that, do not need special consideration.

At the same time, the calculation of $\xi_{ik}(t)$, is fairly complicated and it is therefore desirable to formulate this condition without involving these quantities.

Let us introduce the following notation. Let t be fixed ($t = 1, 2, \ldots, T-1$), and let $l(t)$ be the index for which the following inequalities hold:

$$t + l(t) \le T,$$

$$a_t + a_{t+1} + \cdots + a_{t+l(t)} \ge 4,$$

$$a_t + a_{t+1} + \cdots + a_{t+l(t)-1} < 4.$$

If no such $l(t)$ exists, then let $l(t) = T - t$. According to this definition, $l(t)$ represents the number of periods starting with period t whose total length is at least 4 h, provided that these 4 h following period t (including the a_t length of period t) belong to the planning interval. If this condition is not fulfilled, then $l(t)$ is the number of periods of the planning interval following the tth period. The reason for introducing this notation is that in the subdivision of the planning interval there are periods whose length is not equal to 1 h. On the other hand, according to the requirement, the shutoff units must be in standstill for at least 4 h.

Using this notation, the minimal 4-h standstill requirement can be formulated as follows. For all $i = 1, 2, \ldots, K$, $k = 1, 2, \ldots, N(i)$ and $t = 1, 2, \ldots, T-1$ the following implication holds: whenever both

$$\sum_{j:k\in J(i,j)} y_{ij}^{t-1} = 1 \quad \text{and} \quad \sum_{j:k\in J(i,j)} y_{ij}^{t} = 0$$

hold,

$$\sum_{j:k\in J(i,j)} y_{ij}^{t+l} = 0$$

must hold for all $l = 1, 2, \ldots, l(t)$, too.

It can easily be seen that the preceding relation is enforced by prescribing the following system of inequalities:

$$\sum_{j:k\in J(i,j)} y_{ij}^{t-1} - \sum_{j:k\in J(i,j)} y_{ij}^{t} + \sum_{j:k\in J(i,j)} y_{ij}^{t+l} \le 1, \qquad (2.3.26)$$

$$t = 1, 2, \ldots, T - 1,$$
$$l = 1, 2, \ldots, l(t),$$
$$i = 1, 2, \ldots, K,$$
$$k = 1, 2, \ldots, N(i).$$

These constraints require that if a unit is shut off in a period corresponding to the planning time interval, it should not be restarted within the following 4 h. It remains to ensure that those power plant units that have been shut off within the last 4 h of the previous planning stage should in the current planning stage only be available for startup after a standstill of at least 4 h. These types of start-and-stop constraints will be called *first-startup constraints*.

To formulate them, let l_{ik} be the number of unchanged inoperative periods needed for unit k at power plant i. That is, in the case of $\sum_{j:k \in J(i,j)} y_{ij}^0 = 0$ and $\xi_{ik}(0) < 4$, let l_{ik} be the serial number of the period where

$$\xi_{ik}(0) + a_1 + \cdots + a_{l_{ik}} \geq 4$$

and

$$\xi_{ik}(0) + a_1 + \cdots + a_{l_{ik}-1} < 4$$

hold; otherwise let $l_{ik} = 0$.

For all the units of every power plant we require that

$$\sum_{j:k \in J(i,j)} y_{ij}^t = 0$$

must hold in the first l_{ik} periods of the current planning stage. Therefore, the first-startup constraints are as follows:

$$\sum_{j:k \in J(i,j)} y_{ij}^l = 0, \quad i = 1, 2, \ldots, K, \quad l = 1, 2, \ldots, l_{ik}, \quad k = 1, 2, \ldots, N(i).$$

$$(2.3.27)$$

Recall that, according to the remark in Sect. 2.1.1, further individual restrictions may be necessary with respect to the mode-of-operation variables. It may also happen that for some reason, certain modes of operation cannot function consecutively at certain power plants. These subjects will not be discussed further in this book.

Figure 2.4 displays the structure of stop-and-start constraints using a matrix pattern (the first-startup constraints are neglected). The nonzero blocks in the figure, indicated as unit blocks, are in reality summation matrices, with their structure displayed separately in Fig. 2.5.

Fuel constraints also establish connections among the variables of separate periods.

According to the discussion in Sect. 1.2.2, to provide faultless operation of an electric power system, it may be necessary to restrict the consumption of primary

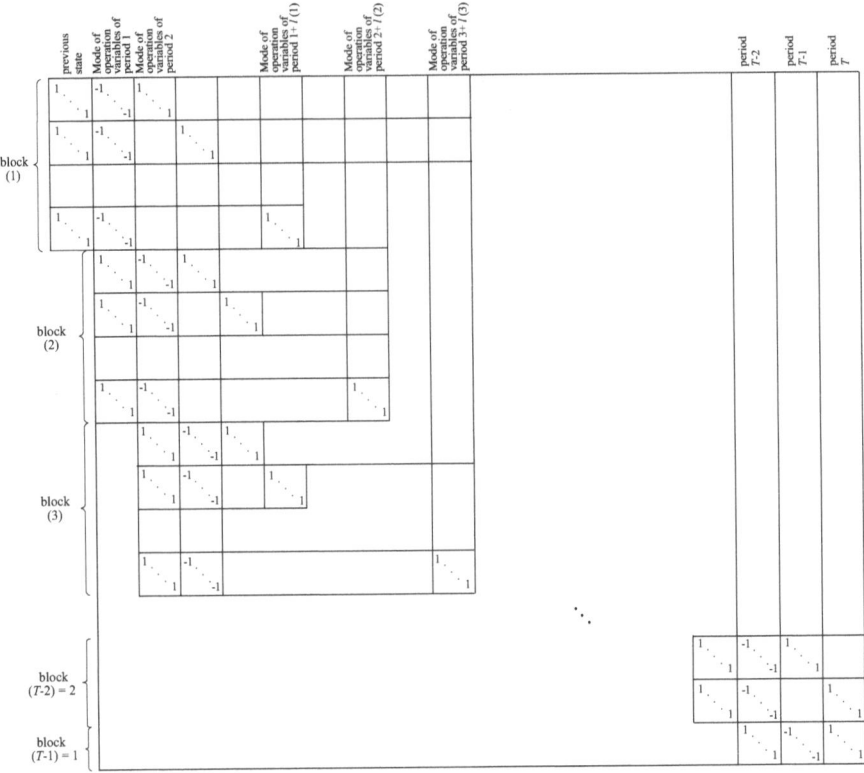

Fig. 2.4 Schematic overview of stop-and-start constraints

energy sources of a power plant (or plants). This restriction may mean a lower or an upper bound, possibly a lower and an upper bound, or even an equality constraint.

In the model, restriction of the primary energy sources of a power plant in the planning stage is accomplished by including fuel constraints. Power plants where such constraints are necessary will be called *power plants with fuel constraints*.

Let i be the index of a power plant with fuel constraints. The total production of power plant i in the planning stage can be expressed by the sum

$$\sum_{t=1}^{T} a_t \sum_{j=1}^{M(i)} (P_{ij}^{\min} y_{ij}^t + p_{ij}^t).$$ (2.3.28)

The quantity of the generated electric power is directly related to the quantity of the consumed primary energy source. Therefore, bounds on the primary energy source can easily be transformed into bounds on the quantity of power to be generated. The fuel constraints are formulated as

$$R_{i\,\min} \le \sum_{t=1}^{T} a_t \sum_{j=1}^{M(i)} (P_{ij}^{\min} y_{ij}^t + p_{ij}^t) \le R_{i\,\max},$$ (2.3.29)

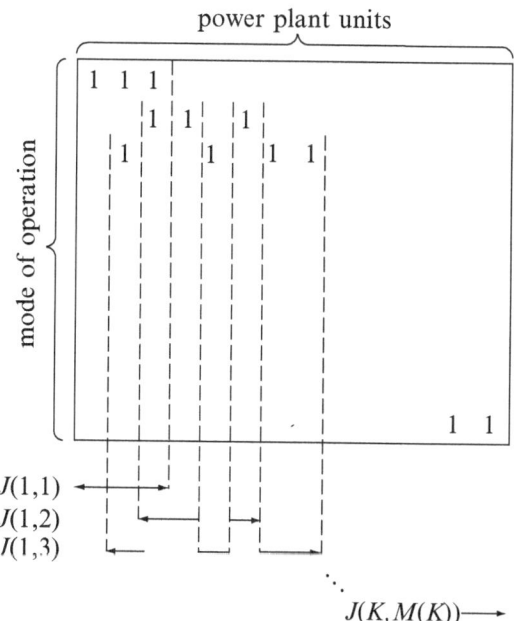

Fig. 2.5 Structure of a summation block in stop-and-start constraints

where $R_{i\,\min}$ and $R_{i\,\max}$ represent respectively the prescribed lower and upper bounds on the amount of electric energy to be generated.

Let us note that the restrictions as formulated previously refer to the electric power generated in the whole planning stage – $\left(\sum_{t=1}^{T} a_t \text{ hours}\right)$ – while in practice consumption of the primary energy source in a 24-h period is usually restricted. In cases where the values of $R_{i\,\min}$ and $R_{i\,\max}$ represent restrictions for 24 h, the generated power should only be added up for the periods corresponding to the given 24-h time interval.

2.4 Model Structure, Size, and Characteristics

In this final subsection concerning the general model, we briefly discuss the structure and summarize the main characteristics of the nonlinear mixed integer programming problem corresponding to the general model.

The variables of the model are the mode-of-operation variables ($\mathbf{y}$), the production-level variables ($\mathbf{p}$), and the voltage variables ($\mathbf{v}$, $\mathbf{w}$). The components of the mode-of-operation vector variable can only have a value of 0 or 1, and the production-level variables are nonnegative.

The objective function to be minimized is the sum of the following three terms. The production-cost part (2.2.4) of the objective function is a separable function of the y_{ij}^t and p_{ij}^t variables that is linear in the mode-of-operation variables $\mathbf{y}^t$ and nonlinear but convex in the production-level variables $\mathbf{p}^t$. The cost of standstill (2.2.8) is the sum of nonlinear functions of the components y_{ij}^t, $t = 1, 2, \ldots, T$, corresponding to the modes of operation of the individual power plants. The third term (2.2.10) represents the costs due to transmission losses; it is a periodwise convex quadratic function of the $\mathbf{v}^t$, $\mathbf{w}^t$ variables. Thus, for fixed values of the mode-of-operation variables the objective function is convex and separable in the variables $\mathbf{p}$, $\mathbf{v}$, and $\mathbf{w}$.

The system of constraints is structured. To each of the periods correspond constraints involving only the variables of that period. Thus, for a fixed period, for the components of the mode-of-operation variables the SOS constraints (2.1.1) must hold, while between $\mathbf{y}^t$ and $\mathbf{p}^t$ the linear (2.1.2) variable coupling constraints provide connections. The supply constraints (2.3.14) are equality constraints and include all the variables $\mathbf{y}^t$, $\mathbf{p}^t$, $\mathbf{v}^t$, and $\mathbf{w}^t$. They are nonlinear equality constraints since on the right-hand side a nonlinear function of $\sum_{j=1}^{M(i)} (P_{ij}^{\min} y_{ij}^t + p_{ij}^t)$ and a nonlinear transmission loss function appear. The latter function is a convex quadratic function of the $\mathbf{v}^t$, $\mathbf{w}^t$ variables.

Network constraints are also constraints repeated for periods. Power balance constraints (2.3.20), connecting the variables $\mathbf{y}^t$ and $\mathbf{p}^t$ with the node-voltage variables $\mathbf{v}^t$, $\mathbf{w}^t$, are nonlinear equality constraints. The voltage limit constraints (2.3.21) are inequality constraints prescribing lower and upper bounds on the absolute value of the voltages at the nodes. Although the absolute value function is a convex function of v_i and w_i, prescribing lower bounds results in nonconvex constraints. The branch-load constraints (2.3.22) are inequality constraints involving nonlinear and nonconvex functions of $\mathbf{v}^t$ and $\mathbf{w}^t$. Finally, the reactive power injection constraints (2.3.24) are nonlinear inequality constraints, including the variables $\mathbf{y}^t$, $\mathbf{v}^t$ and $\mathbf{w}^t$. They are linear in the variables $\mathbf{y}^t$ but nonlinear and nonconvex in the variables $\mathbf{v}^t$ and $\mathbf{w}^t$.

The constraints connecting the periods are as follows. The linear stop-and-start constraints, (2.3.26) and (2.3.27), include the mode-of-operation variables of several periods. Their structure can be seen in Fig. 2.4. Regarding the fuel constraints (2.3.29), each particular constraint contains the variables of several periods. Nevertheless, it is a linear function of the mode of operation and the production-level variables belonging to just one power plant.

Figure 2.6 shows a schematic sketch of the model, where the production level, mode-of-operation, and voltage variables of a given period are given consecutively.

Fixing the values of the mode-of-operation variables y_{ij}^t results in a nonlinear programming problem with continuous variables. As discussed earlier, the objective function to be minimized is convex. However, with the exception of the fuel constraints and the upper bounding side of the voltage limit constraints, the constraints are nonconvex from the point of view of mathematical programming. The reason is that those constraints either are of the nonlinear equality type or involve nonconvex functions.

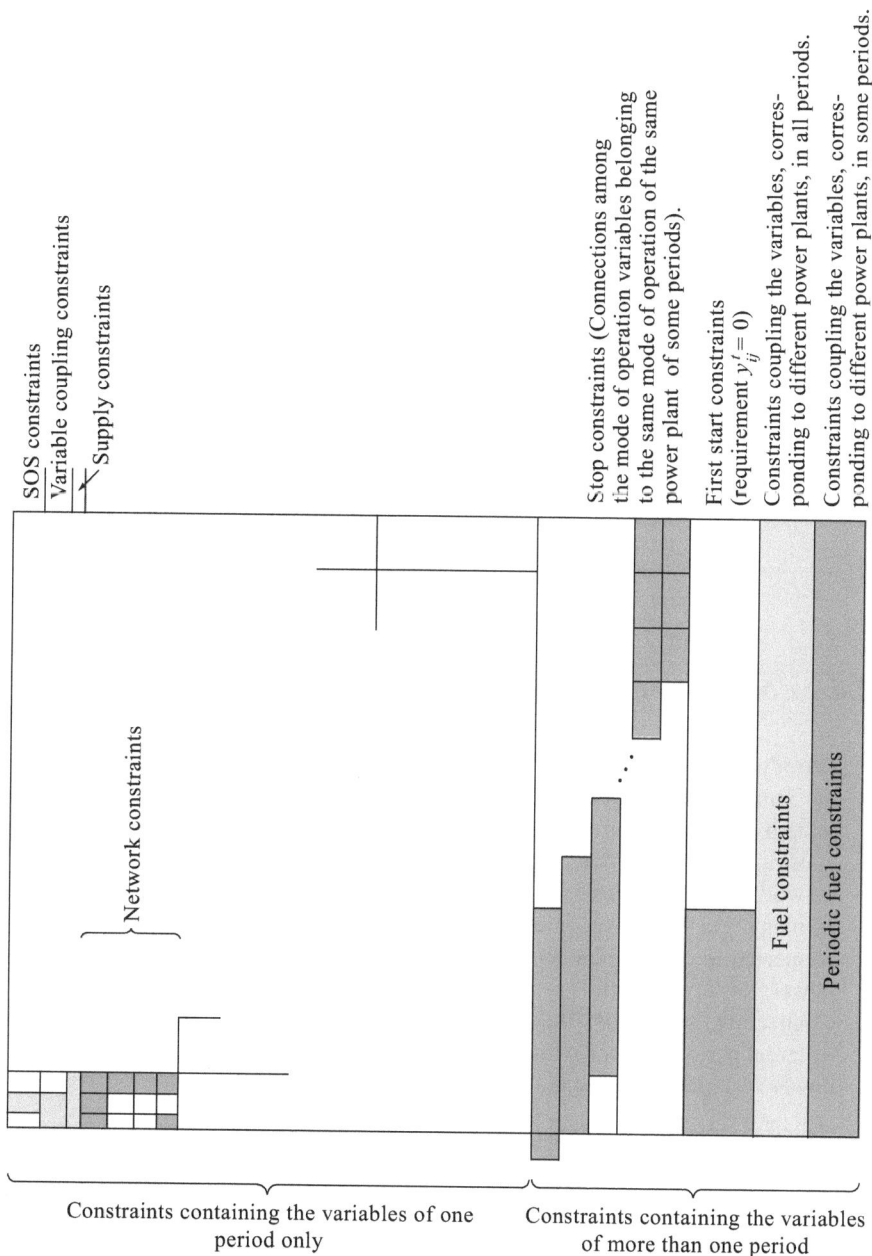

Fig. 2.6 Structure of general model

The size of the nonlinear mixed-variable optimization problem, corresponding to the general model, is as follows.

2.4.1   Number of Model Variables

1. The number of the model's $0-1$ variables $= T \times$ the number of the available modes of operation in the electric power generating system:

$$T \sum_{i=1}^{K} M(i).$$

2. The number of production-level variables $= T \times$ the number of the modes of operation available in the electric power generating system:

$$T \sum_{i=1}^{K} M(i).$$

3. The number of continuous voltage variables $= T \times$ the number of network nodes:

$$T \times N.$$

2.4.2   Number of Model Constraints

1. Periodwise: the number of SOS constraints equals the number of power plants; the number of modes of operation available in the electric power generating system equals the number of variable coupling constraints; there is one supply constraint; there are twice as many voltage limit constraints as the number of network nodes, twice as many branch-load constraints as the number of branches, as many reactive power injection constraints as the number of network nodes connected to controllable sources of reactive power. Additionally, there are the power flow constraints.
2. Connecting the periods: The number of start-and-stop constraints connecting the different periods is approximately $4 \times T$ times the number of units in the electric power system; furthermore, there are at most five fuel constraints.

Note that if an *enumeration type strategy* were used to solve a problem, i.e., the problem were solved using fixed values of the mode-of-operation variables, then the size of the problem to be solved in the individual iterations would be as follows. Number of variables: $T(K + N)$.
Number of constraints:

1. Periodwise: as many lower and upper bounds on the production-level variables as the number of power plants (they correspond to the variable coupling constraints), one supply constraint, and the same number of network constraints as before.

2. In this problem, just the fuel constraints would connect the periods. There would be five of them at most.

If there is no fuel constraint in a specific planning stage, the problem decomposes into T independent separate problems (because the mode-of-operation variables are fixed!). Each of them contains as many production-level variables as the number of power plants and twice as many voltage variables as the number of transmission network nodes. Their objective function is a convex, separable, nonlinear function of the production-level variables. There are some nonlinear constraints as well. For example, the constraints describing network conditions are quadratic, but not necessarily convex.

2.5 Summary of Notations in Chap. 2

T	Number of periods in planning stage;
a_t	Length of tth period in hours;
K	Number of power plants;
$M(k)$	Number of modes of operation available in power plants in considered planning stage, $k = 1, 2, \ldots, K$;
i, j	Pair of indices referring to jth mode of operation of ith power plant, $i = 1, 2, \ldots, K$, $j = 1, 2, \ldots, M(k)$;
$\mathbf{y}^t, y_l^t, y_{ij}^t$	Vector of mode-of-operation variables and its components, corresponding to tth period;
$\mathbf{y}$	Vector of all mode-of-operation variables (formed by concatenating vectors $\mathbf{y}^t$);
$\mathbf{y}^0$	Value of vector of modes of operation at last period preceding current planning stage;
$\mathbf{p}^t, p_l^t, p_{ij}^t$	Production-level variable vector and its components corresponding to tth period;
$\mathbf{p}$	Vector of all production-level variables in model (formed by concatenating vectors $\mathbf{p}^t$);
$P_{ij}^{\min}, P_{ij}^{\max}$	Lower and upper bounds on jth mode of operation at ith power plant;
$\mathbf{v}^t, \mathbf{w}^t$	Vectors formed by real and imaginary parts of complex potential at nodes of network in tth period;
$f_{ij}(P)$	Production costs as function of produced power for jth mode of operation at ith power plant;
$K_{ij} = f_{ij}(P_{ij}^{\min})$	Cost of production on minimal level for jth mode of operation at ith power plant;

$k_{ij}(P) = f_{ij}(P_{ij}^{\min} + P)$ $- f_{ij}(P_{ij}^{\min})$	Production costs of excess power P above minimum level corresponding to jth mode of operation at ith power plant;
$N(i)$	Number of units at ith power plant;
$J(i, j)$	Set of serial numbers of units operating during jth mode of operation at ith power plant;
$g_{ik}(\tau)$	Cost of standstill of unit k at power plant i as function of standstill time;
$G_{ik}(\infty), G_{ik}(0), c_{ik}$	Constants appearing in function of cost of standstill;
$\tau_{ij}(t)$	Length of continuous standstill of jth mode of operation at ith power plant up to period t;
$\tau_{ij}(0)$	Length of continuous standstill preceding current planning stage for jth mode of operation at ith power plant;
$\xi_{ik}(t)$	Length of continuous standstill of kth unit at ith power plant up to tth period;
$\xi_{ik}(0)$	Length of continuous standstill of kth unit preceding current planning stage at ith power plant;
$P_{ij}^{\text{self}}(P)$	Amount of self-consumption during production on level P in jth mode of operation at ith power plant;
$P^{t\,\text{loss}}$	Transmission losses in period t (a function of $\mathbf{v}^t, \mathbf{w}^t$);
$P^{t\,\text{self}}$	Total amount of self-consumption of whole electric power system in period t (a function of $\mathbf{y}^t, \mathbf{p}^t$);
$P^{t\,\text{dem}}$	Power demand in tth period, with the value (MW) of the demand curve corresponding to tth period;
$l(t)$	Number of periods following period t whose total length – including length of period t – is at least 4 h in duration, respectively number of periods following period t in current planning stage;
l_{ik}	The kth unit at ith power plant can only be started following first l_{ik} periods of planning stage;
$R_{i\,\min}, R_{i\,\max}$	Bounds on production imposed by fuel constraints for power plants with fuel constraints;
N	The number of the nodes of the network;
M	Number of branches of network;
I_E	Index set consisting of serial numbers of nodes to which power plants are connected;
N_E	Number of elements of I_E;
I_M	Set of serial numbers of nodes with connected controllable source of reactive power;
N_M	Number of elements of I_M;
P_i	Difference between amounts of active power injection and consumption at ith node, $i = 1, \ldots, N$;

Q_i	Difference between amounts of reactive power injection and consumption at ith node, $i = 1, \ldots, N$;
P_i^G	Active power injection at ith node, $i = 1, \ldots, N$;
P_i^F	Active power consumption at ith node, $i = 1, \ldots, N$;
$\hat{\mathbf{P}}_i^G$	N_E-dimensional vector formed by $i \in I_E$ components of vector $\mathbf{P}^G$;
$\mathbf{D}$	Matrix of linear transformation mapping production-level and mode-of-operation variables into generated power at nodes;
$\mathbf{p}^{\min}$	Its components, corresponding to power plants and modes of operation, are the minimal production levels;
v_i	Real part of voltage at ith node;
w_i	Imaginary part of voltage at ith node;
$V_i^{\min}$	Lower bound on absolute value of voltage at ith node;
$V_i^{\max}$	Upper bound on absolute value of voltage of ith node;
$Q_i^{\min}(\mathbf{y})$	Lower bound on reactive power, $i \in I_M \cap I_E$;
$Q_i^{\max}(\mathbf{y})$	Upper bound on reactive power, $i \in I_M \cap I_E$;
$Q_i^{\min}$	Lower bound on reactive power, $i \in I_M \setminus I_E$;
$Q_i^{\max}$	Upper bound on reactive power, $i \in I_M \setminus I_E$;
$T_{ik}^{\max}$	Upper bound on apparent power related to branch (i, k) (thermic loadability);
$f_i(\mathbf{v}, \mathbf{w})$	Function describing dependence of active power corresponding to node i on network voltage distribution;
$g_i(\mathbf{v}, \mathbf{w})$	Function describing dependence of reactive power corresponding to node i on network voltage distribution;
$S_{ik}(v_i, v_k, w_i, w_k)$	Power flowing out from ith node into branch (i, k);
$P^v(\mathbf{v}, \mathbf{w})$	Active power loss as function of voltage distribution.

Chapter 3
Assumptions for Model Simplification

In Chap. 2, our aim in formulating the general model of the scheduling problem was to construct a model that would best describe the problem, regardless of whether we had any chance of solving the corresponding mathematical programming problem. According to the overview of the model in Sect. 2.4, the corresponding problem is a large-scale, mixed-variable problem containing nonlinearities both in its objective function and constraints. Due to its size and complexity, there is no way to solve realistically sized instances of this problem numerically.

An attempt should be made to decrease the number of variables and constraints and eliminate the nonlinearities in such a way that the model should *still be* a proper model of the scheduling problem and the solution of the corresponding mathematical programming problem should be an *already* manageable task.

This chapter discusses simplifying assumptions that will help to simplify the general model in the foregoing sense.

To model the scheduling problem, it is *advisable* to consider these assumptions since the mathematical programming problem that arises in this way will be numerically tractable.

Our assumptions are well founded because they are based on characteristics of long-standing technologies of operations management or employ mathematical approximation techniques. There are also some simplifications with respect to the applied notations.

The next chapter, Chap. 4, includes a description of the simplified model. The applied planning stage corresponds to a 25-h time interval and contains $T = 27$ periods, $a_t = 1.0$ or 0.5, $t = 1, 2, \ldots, 27$.

A. Prékopa et al., *Scheduling of Power Generation*, Springer Series in Operations Research and Financial Engineering, DOI 10.1007/978-3-319-07815-1_3, © Springer International Publishing Switzerland 2014

3.1 Simplifying Assumptions Based on the Characteristic Shape of the Demand Curve

A characteristic feature of the demand curve, which provides information about daily electric power demand, is that between pairs of two maximal values of the power demand there are several subsequent periods where the demand monotonically decreases, followed by some periods in which it varies to a small extent around a minimal value and finally a few periods follow with monotonically increasing demand (see Fig. 1.1 on p. 7).

It seems reasonable to require that in time intervals with monotonically decreasing demand no changeovers should be allowed that would require the startup (i.e., heating up) of power plant units, that is, only those changeovers should be permitted that involve shutting down some power plant units.

In time intervals where the demand increases monotonically, only those changeovers will be permitted that originate from an already active mode of operation by starting up additional units.

In those time intervals where demand changes only slightly, no changeovers will be permitted; only the production levels of active modes of operation may be varied.

Based on the preceding discussion, as a simplifying assumption, a planning stage of 1 day can be subdivided into six subsequent phases.

The first and fourth phases are characterized by monotonically decreasing demand. In the third and sixth phases, demand increases in a monotonous way. The second and fifth phases should contain four periods each, where the demand fluctuates around a minimal value.

Let us call the first and fourth phases *shutdown phases*, the third and sixth phases *startup phases*, and the second and fourth phases *phases of stagnation*.

In the search for an optimum let us confine ourselves to those schedules where in the shutdown phases only changeovers that involve merely shutdowns of units are permitted, in the startup phases only changeovers that can be attained by startups are considered as feasible, and finally, in phases of stagnation there are no changeovers.

This restriction of the number of possible schedules is in accordance with the practice of operations management of power plants and follows in a natural way from information related to the cost impacts of changeovers.

At the same time, from the point of view of the model, this is a simplifying assumption because schedules complying with the assumption automatically ensure a standstill of the shutoff equipment for a minimum of 4 h. Therefore, the stop-and-start constraints described in Sect. 2.3.2 can be dropped from the model, resulting in a decrease in the number of constraints.

Note that the regulations concerning operations management in power plants are roughly similar to those in Hungary, with minor differences [46].

3.2 Specification for Ordering the Mode of Operations

The comment in the final part of Sect. 2.2.1 shows that in most cases the modes of operation that are feasible in a given planning stage can be ordered in such a way that

$$J(i, j_1) \supset J(i, j_2) \supset \cdots \supset J(i, j_{M(i)})$$

holds for the sets $J(i, j)$, $i = 1, 2, \ldots, K$, $j = 1, 2, \ldots, M(i)$, consisting of the serial numbers of the corresponding power plant units.

The simplified model is formulated for a planning stage where the modes of operation have the aforementioned characteristics. It is also supposed that for the order of the modes of operation

$$J(i, 1) \supset J(i, 2) \supset \cdots \supset J(i, M(i))$$

holds for all $i = 1, 2, \ldots, K$.

In other words, a changeover to a mode of operation of a higher serial number can be carried out by shutting off some power plant units.

On the one hand, this simplifying assumption makes it easier to include in the model those features described in Sect. 3.1; on the other hand, it facilitates the formulation of the costs of standstill of the power plant units. That is, if $J(i, 1) \supset J(i, 2) \supset \cdots \supset J(i, M(i))$ holds, then those units for which $k \in J(i, l) \backslash J(i, l + 1)$ is satisfied are in standstill in periods when neither mode of operation l nor a mode of operation preceding this mode of operation is active. Here and in the sequel, the term *mode of operation preceding mode of operation j* respectively *mode of operation following mode of operation j* is used, meaning that in the order of the modes of operation a particular mode of operation precedes respectively follows mode of operation j, i.e., its serial number is smaller respectively larger.

The aforementioned relation between the standstill of the mode of operation and of the power plant units makes it possible that subsequently the costs of standstill of the modes of operation will be considered instead of the costs of standstill of the power plant units.

Let $s_{ij}(\tau)$ be the function of the costs of standstill of the jth mode of operation in the ith power plant. The value of $s_{ij}(\tau)$ is defined as

$$s_{ij}(\tau) = \sum_{k \in J(i,j) \backslash J(i,j+1)} g_{ik}(\tau), \quad i = 1, 2, \ldots, K, \quad j = 1, 2, \ldots, M(i) - 1.$$

The costs of standstill of power plant units of serial number $k \in J(i, j) \backslash J(i, j + 1)$ must be taken into account in those periods when a mode of operation following the jth mode of operation is active. Therefore, the $s_{ij}(\tau)$ costs of standstill of the jth mode of operation should also be considered in those periods when one of the modes of operation following the jth mode of operation is active.

3.3   Approximation of Production Costs

According to Sect. 2.2.1, the partial costs of a P level production of the jth mode of operation at the ith power plant can be specified by the function $f_{ij}(P)$, $i = 1, 2, \ldots, K$, $j = 1, 2, \ldots, M(i)$.

As a simplifying assumption, instead of these nonlinear cost functions $f_{ij}(P)$, let us consider a piecewise-linear approximation of them (Fig. 3.1).

Let the number of line segments of the approximating cost function of the jth mode of operation at the ith power plant be $r(i, j)$.

Let the power values corresponding to the breakpoints of the approximating function be

$$P_{ij}^{\min} = P_{ij1}^{\min},$$

$$P_{ij1}^{\max} = P_{ij2}^{\min},$$

$$P_{ijr(i,j)-1}^{\max} = P_{ijr(i,j)}^{\min},$$

$$P_{ijr(i,j)}^{\max} = P_{ij}^{\max}. \tag{3.3.1}$$

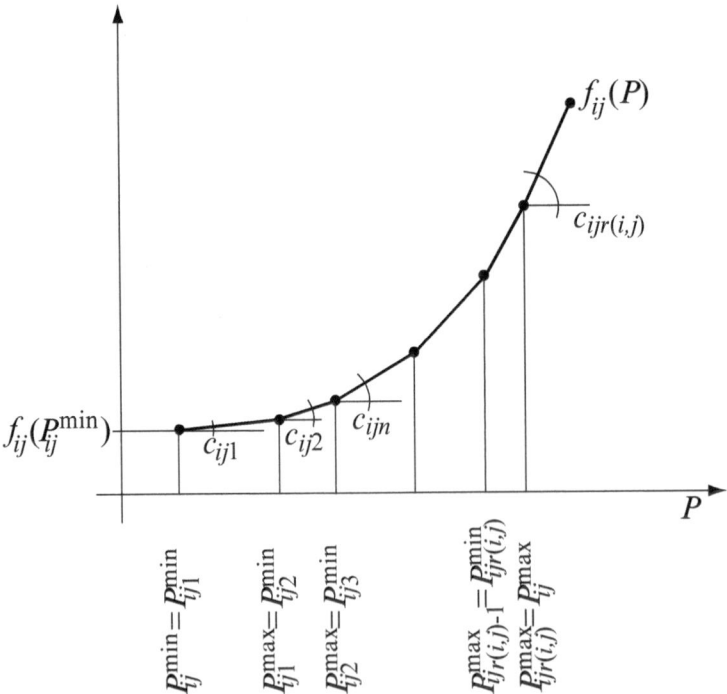

Fig. 3.1 Piecewise-linear approximation of production costs

Let the slopes of the approximating line segments be c_{ijl}, $l = 1, 2, \ldots, r(i, j)$. The cost functions $f_{ij}(P)$ are such that the following relations hold with respect to the slopes:

$$c_{ij1} < c_{ij2} < \cdots < c_{ijr(i,j)}. \qquad (3.3.2)$$

Concerning this approximation, the cost of production level P in the jth mode of operation of the ith power plant is as follows (provided that $P_{ijl_0}^{\min} \le P \le P_{ijl_0}^{\max}$, i.e., P belongs to the l_0th line segment):

$$f_{ij}(P) \sim K_{ij} + \sum_{l=1}^{l_0-1} (P_{ijl}^{\max} - P_{ijl}^{\min})c_{ijl} + (P - P_{ijl_0}^{\min})c_{ijl_0}, \qquad (3.3.3)$$

where $K_{ij} = f_{ij}(P_{ij}^{\min})$ holds, in accordance with the notations introduced in Chap. 2.

If the function of production costs is replaced by a piecewise-linear function as described previously, then the number of production level variables must be increased. Therefore, in the simplified model there are $r(i, j)$ variables instead of a single p_{ij}^t variable of the general model (Sect. 4.1.2).

3.4 Approximation of Changeover Costs

The partial costs due to standstill and changeovers can be specified as described in Sect. 2.2.2. The resulting function, however, is a complicated nonlinear function of the 0–1 variables characterizing the modes of operation.

In this section we will show the assumption that can be used to approximate these costs with a linear function of the already mentioned variables.

To obtain an approximation we will make the following assumption: the shutdowns and startups of the power plant units are carried out in a symmetric way. This assumption means that if a power plant unit is shut off l periods prior to the stagnation phase, then the restart takes place at the end of the lth period following the stagnation phase. (This assumption is closely related to the discussion in Sect. 3.1; it is justified by the specific characteristics of the demand curve.) As a consequence of our assumption, there are no so-called overlapping costs among the individual planning intervals. If a unit was shut off prior to the analyzed planning stage, then its approximate costs of standstill had to be taken into account in the previous planning interval. Similarly, if a power plant unit is shut off in the present planning stage, the costs of standstill are taken into account in the present planning stage.

The assumption of symmetric stop–restart is only made to simplify the computation of the costs of standstill. However, in the model, other schedules, including modes of operation of nonsymmetric stop–restart, are permitted as well.

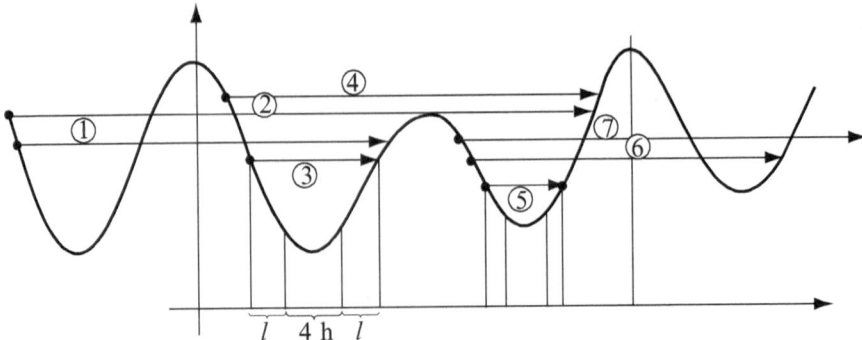

Fig. 3.2 In the case of symmetric stop–restart, stops and restarts indicated by ① ② ④ ⑥ ⑦ cannot occur

The approximate linear function of the cost of changeover is such that if a symmetric stop and restart takes place, the function value is equal to the cost as computed in Sect. 2.2.2; otherwise, it provides an approximation of that cost.

To better understand the notion of symmetric stop and restart, let us consider Fig. 3.2.

Figure 3.2 shows adjoining demand curves and pairs of stops and restarts. The starting points of the arrows indicate the shutdown of a power plant unit, while the endpoints of the arrows indicate a restart of that unit.

	Time of stop	Time of start
1	Previous day 2nd shutdown phase	1st startup phase
2	Previous day 2nd shutdown phase	2nd startup phase
3	1st shutdown phase	1st startup phase
4	1st shutdown phase	2nd startup phase
5	2nd shutdown phase	2nd startup phase
6	2nd shutdown phase	following day 1st startup phase
7	2nd shutdown phase	following day 2nd startup phase

Symmetric restart means that only types ③ and ⑤ may occur, the consequence of which is that the cost has no portion that would overlap with the next day.

In Fig. 3.2 it can be seen quite clearly that the assumption is supported by the characteristic shape of the demand curve (approximately the same demand values are located symmetrically with respect to the phase of stagnation).

In the approximation of the cost of standstill, the cost of a $4 + 2l_0$ standstill associated with stopping and restarting a unit or units (depending on the changeover) shut off l_0 periods prior to the phase of stagnation will be represented as a sum in a way similar to that given in Sect. 2.2.2 (this standstill is due to the 4-h stagnation and the l_0 periods prior to it and following it, resulting in a $2l_0$-h standstill; here the possibility that there are also some 30-min periods is disregarded).

To represent this as a sum, let us define the quantities d_{ij}^t in the following way (for every mode of operation in every power plant and in every period, apart from the second, third, and fourth periods of the stagnation phases):

$$
d_{ij}^t = \begin{cases} s_{ij}(4+2l) - s_{ij}(4+2l-1) & \text{if } t \text{ is the } l\text{th period prior to the stagnation phase,} \\ s_{ij}(4) & \text{if } t \text{ is the beginning of the stagnation phase,} \\ s_{ij}(4+2l-1) - s_{ij}(4+2l-2) & \text{if } t \text{ is the } l\text{th period after the stagnation phase.} \end{cases}
$$

(3.4.4)

Let us assume that the jth mode of operation of the ith power plant is shut off at the beginning of the l_0th period prior to the stagnation phase that starts at t_0 and that its restart is performed in a symmetric way. In this case,

$$
\sum_{t=t_0-l_0}^{t_0} d_{ij}^t + \sum_{t=t_0+4}^{3+t_0+l_0} d_{ij}^t = s_{ij}(4+2l_0)
$$

(3.4.5)

holds, i.e., the sum of the d_{ij}^t quantities over the standstill periods equals the corresponding cost of standstill. The addition of the d_{ij}^t quantities means the addition of the increments of the cost of standstill, like the addition presented in Sect. 2.2.2. The order of addition is different, though. Figure 3.3 highlights this.

If the start does not take place in a symmetric way and the addition is carried out with the actual standstill times, there is a deviation from the value $s_{ij}(4+2l_0)$ consisting of the partial sum, which corresponds to the second term in the addition. We assume that the standstill costs obtained in this way are good-enough approximations of the actual $s_{ij}(\tau)$ costs.

In the model we must ensure that the costs of standstill corresponding to those units that are shut off due to stops and restarts of types ① and ②, according to the notation in Fig. 3.2, should not be charged during the period in question. When a ④ type of stop-restart OCCURS, no standstill costs should be taken into account in the case of equipment that is still in standstill in the second stop–stagnation–start phase. The costs of stop–restart of the ⑥ and ⑦ types are given by an approximation of the costs of symmetric stop–restart, as explained earlier. Stop-restarts of the ③ and ⑤ types are either symmetric approximately symmetric.

3.5 Introduction of an Operating Point and Some Notations

Regarding those parts of the general model presented in Chap. 2 that involve a transmission network, the functions appearing in constraints (2.3.20), (2.3.22), and (2.3.24), as well as the objective function (2.2.10), are nonlinear. Since our

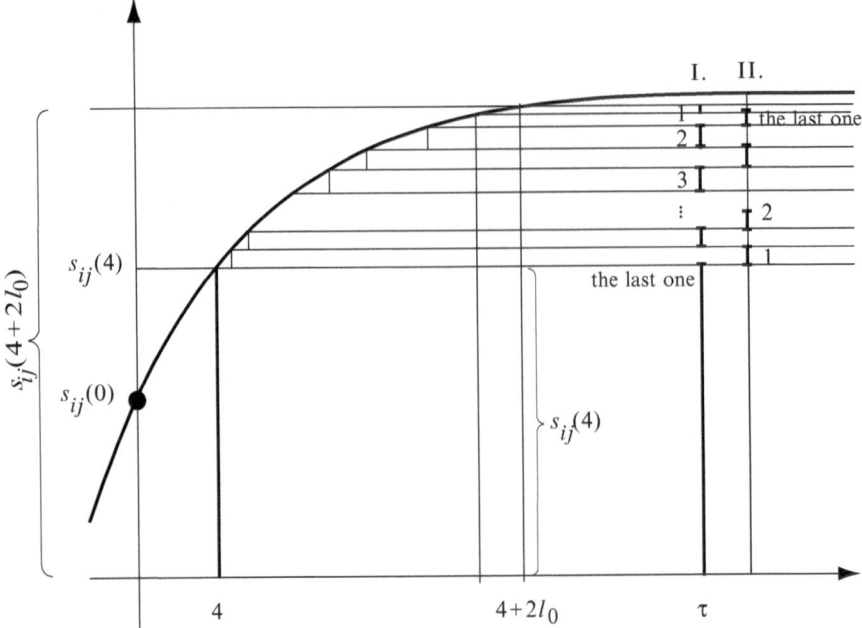

Fig. 3.3 Cost of standstill as a sum of d_{ij}^t increments. Increments indicated by I on the right-hand side provide the first part of the summation; those indicated by II provide the second part of the summation; Arabic numerals indicate the order of the terms

aim is to build a linear model, we wish to approximate these functions by linear functions. The procedure will be as follows: an operating point will be fixed, a linearization around the operating point will be carried out, and by imposing bounds on the variables, it will be ensured that the accuracy of the approximation is sufficiently good in the corresponding neighborhood of the operating point.

Naturally, the operating point will be a point for which the network constraints (A.3.43) hold; therefore, its specification requires the solution of the power flow problem (Sect. A.3). This requires the specification of the absolute values of the voltages at the nodes connected to controllable sources of reactive power (defined in Sect. 2.3.1; see Sect. A.3) and the value of the voltage at the reference node. Regarding the model, these quantities are prescribed input data. Furthermore, the power consumption at the consumers' nodes and the generated active power at the nodes of the power plants must be known. The former are input data while the latter are provided by solving an optimization problem for the particular period with respect to power generation (for details see Sects. 4.3.1 and 5.4).

For a detailed discussion of the process of linearization, some notations must be introduced. In the sequel, the superscript t, indicating the periods for the variables, will be omitted. Let $(\mathbf{v}^*, \mathbf{w}^*)$ denote the operating point. In contrast to the notations in Sect. 2.3.1, $L(= N_M)$ denotes the number of nodes connected to controllable sources of reactive power. According to the discussion in Sect. A.3, the set of nodes connected to controllable sources of reactive power consist of the (Q, V) nodes and contains the reference node. The rest of the nodes will be called consumer nodes, and their set consists of the (P, Q) nodes according to Sect. A.3. Let N_F denote the number of consumer nodes. Vector $\mathbf{v}$ will be partitioned in the following way:

$$\mathbf{v} = \begin{pmatrix} \mathbf{v}^M \\ \mathbf{v}^F \end{pmatrix}, \tag{3.5.6}$$

where $\mathbf{v}^M \in R^L$ corresponds to nodes connected to controllable sources of reactive power and $\mathbf{v}^F \in R^{N_F}$ corresponds to consumer nodes. It is assumed that the node serial numbers start with nodes having attached controllable sources of reactive power. Note that the superscript F is used in two different contexts. It indicates consumer nodes in the case of voltage, whereas in the case of active power it indicates power consumption (demand).

Let $\mathbf{g}^T(\mathbf{v}, \mathbf{w}) = (g_1(\mathbf{v}, \mathbf{w}), \ldots, g_N(\mathbf{v}, \mathbf{w}))$. This is partitioned according to the components of controllable sources of reactive power and the components of the consumers:

$$\begin{aligned} (\mathbf{g}^M(\mathbf{v}, \mathbf{w}))^T &= (g_1(\mathbf{v}, \mathbf{w}), \ldots, g_L(\mathbf{v}, \mathbf{w})), \\ (\mathbf{g}^F(\mathbf{v}, \mathbf{w}))^T &= (g_{L+1}(\mathbf{v}, \mathbf{w}), \ldots, g_N(\mathbf{v}, \mathbf{w})). \end{aligned} \tag{3.5.7}$$

On the right-hand side in relation (2.3.15), $\mathbf{Q} \in R^N$ is partitioned in a similar way:

$$(\mathbf{Q}^M)^T = (Q_1, \ldots, Q_L); \qquad (\mathbf{Q}^F)^T = (Q_{L+1}, \ldots, Q_N). \tag{3.5.8}$$

With these notations the part of relations (2.3.15) that corresponds to reactive power takes the following form:

$$\mathbf{g}(\mathbf{v}, \mathbf{w}) = Q,$$

respectively

$$\begin{aligned} \mathbf{g}^M(\mathbf{v}, \mathbf{w}) &= \mathbf{Q}^M, \\ \mathbf{g}^F(\mathbf{v}, \mathbf{w}) &= \mathbf{Q}^F. \end{aligned} \tag{3.5.9}$$

The *Jacobi matrix* of the mapping $\mathbf{g}(\mathbf{v}, \mathbf{w}^*): R^N \to R^N$ will be needed at the point $\mathbf{v} = \mathbf{v}^*$. It is denoted by $\mathbf{Y}$ and partitioned in accordance with the nodes connected to controllable sources of reactive power and consumers:

$$\mathbf{Y} = \begin{pmatrix} \mathbf{Y}_1 & \mathbf{Y}_2 \\ \mathbf{Y}_3 & \mathbf{Y}_4 \end{pmatrix} \begin{matrix} \} L \\ \} N_F. \end{matrix} \tag{3.5.10}$$
$$\underbrace{}_{L} \underbrace{}_{N_F}$$

On the basis of (A.3.35) and (A.3.39), the elements of matrix $\mathbf{Y}$ are as follows:

$$Y_{ii} = \sum_{k \in J(i)} w_k^* G_{ik} + v_i^* \sum_{k \in J(i)} (2B_{ik} - \omega C_{ik}) - \sum_{k \in J(i)} v_k^* B_{ik}, \qquad i = 1, \dots, N;$$

$$\tag{3.5.11}$$

$$Y_{ik} = \begin{cases} -w_i^* G_{ik} - v_i^* B_{ik} & \text{if } (i,k) \text{ is a branch of the network,} \\ 0 & \text{otherwise.} \end{cases} \tag{3.5.12}$$

3.6   Reduction in Number of Voltage Variables

The real and imaginary parts of the complex voltage of all the nodes are considered variables in the model from Chap. 2. In this section, an approximate linear relation is constructed between $\mathbf{v}^M$ and $\mathbf{v}^F$, and using this relation $\mathbf{v}^F$ can be expressed as a function of $\mathbf{v}^M$. Therefore, in the simplified model it will be sufficient to regard $\mathbf{v}^M$ as a variable.

In Sect. A.4 it is made clear that the reactive power flow depends mainly on the absolute values of the voltages at the nodes. Regarding the partition of voltage into real and imaginary parts, it is an empirical fact that the reactive power flow depends mainly on the real part and only to a smaller extent on the imaginary part. Like the reasoning used in Sect. A.4, this fact can be supported by heuristic arguments as well. Consequently, in connection with (3.5.9), the imaginary part will be fixed to the value $\mathbf{w}^*$ of the operating point. In this way the following relation arises:

$$\mathbf{g}^F(\mathbf{v}^M, \mathbf{v}^F, \mathbf{w}^*) = \mathbf{Q}^F. \tag{3.6.13}$$

On the basis of (3.5.10), the *Jacobi matrix* of the preceding system of equations is $(\mathbf{Y}_3, \mathbf{Y}_4)$. As a consequence of the connectedness properties of the electric network (as a graph), matrix $\mathbf{Y}_4$ is nonsingular. Therefore, in a neighborhood of $(\mathbf{v}^{*M}, \mathbf{v}^{*F})$, the system of Eqs. (3.6.13) defines $\mathbf{v}^F$ in an implicit way as a function of $\mathbf{v}^M$, i.e., there exists a mapping $\mathbf{h} \colon R^M \to R^{N_F}$ such that $\mathbf{v}^F = \mathbf{h}(\mathbf{v}^M)$ holds. According to the theorem on the derivative of implicit functions, we obtain

$$\frac{\mathbf{dh}(\mathbf{v}^{*M})}{\mathbf{dv}^M} = -\mathbf{Y}_4^{-1}\mathbf{Y}_3. \tag{3.6.14}$$

Neglecting the higher-order terms in the Taylor series around $\mathbf{v}^{*M}$ of the function $\mathbf{h}(\mathbf{v}^M)$, in a sufficiently small neighborhood of $\mathbf{v}^{*M}$ the following linear relation holds approximately:

$$\mathbf{v}^F - \mathbf{v}^{*F} \sim -\mathbf{Y}_4^{-1}\mathbf{Y}_3(\mathbf{v}^M - \mathbf{v}^{*M}). \tag{3.6.15}$$

An approximately valid equality is denoted by "$\sim$" both here and in the sequel.

3.7 Expressing Imaginary Part of Voltages by Active Power Injection; Further Reduction in the Number of Voltage Variables

In this section we will show that, by virtue of linearization, the imaginary parts of the voltages at the nodes can be expressed in terms of the generated power $\mathbf{P}^G$ at the nodes. Since the latter can be expressed in a linear way using the variables of the model to be formulated, in the simplified model, $\mathbf{w}$ might not act as a variable.

Here again, a linearization procedure, based on implicit functions and already applied in Sect. 3.6, could be used. Nevertheless, a different method will be used based on the linearization of power flow relations, which is needed anyway for the branch-load constraints.

It is known that active power flow is mainly sensitive to phase angles. If the voltages are partitioned into real and imaginary parts, it can be shown analogously that the active power flow is mainly sensitive to the imaginary part of voltages. For this reason, the real part of the voltage is fixed at the $\mathbf{v}^*$ value corresponding to the operating point and the power flow is analyzed as a function of $\mathbf{w}$ in a small neighborhood of $\mathbf{w}^*$. To carry out the linearization, let us recall relation (A.4.32), which describes the active power flowing out of the ith node into the branch (i, k):

$$T_{ik}(v_i^*, v_k^*, w_i, w_k) = G_{ik}[v_i^*(v_i^* - v_k^*) + w_i(w_i - w_k)] + B_{ik}[w_i v_k^* - w_k v_i^*]. \quad (3.7.16)$$

According to A.4, $G_{ik} \ll B_{ik}$ holds, and in relative units (see A.4), the deviations between v_i^* and v_k^* and between w_i and w_k are small. Therefore, the following approximation, which is well-established in practice, will be used:

$$T_{ik}(v_i^*, v_k^*, w_i, w_k) \sim B_{ik}(w_i v_k^* - w_k v_i^*). \quad (3.7.17)$$

This relation is formulated for all branches, which can be performed in a compact way by utilizing the following notations.

Let V be an $M \times N$ matrix resulting from the transposed $\mathbf{A}^T$ of the node-edge incidence matrix $\mathbf{A}$ in the following way. Proceeding in a row-by-row fashion in matrix $\mathbf{A}^T$, the real part of the reference point voltage corresponding to the endpoint of the edge is placed in the position of the starting point of the edge, while (-1)-times the amount of the real part of the voltage at the starting point is placed at the position corresponding to the endpoint of the edge. Let $\mathbf{T}$ ($\mathbf{T} \in R^M$) denote the vector of active powers flowing out into the branch from the starting point of the branch. Let $\mathbf{B}$ be a diagonal matrix of size $M \times M$; if the lth branch is (i, k), then let the lth element of the diagonal be $B_{ik}, l = 1, \ldots, M$.

The approximation of the power flow with our notations is as follows:

$$\mathbf{T} \sim \mathbf{B}\mathbf{V}\mathbf{w}. \quad (3.7.18)$$

Let us note that relations (3.7.18) correspond edgewise to the direction of edges as oriented in the directed graph, i.e., they describe the power flowing out into the branches from the starting point of the branch in the directed graph. If the actual power flow is of the direction $k \rightarrow i$, then the power flowing out from i into the branch has a negative value.

In accordance with Sect. 2.3.1, the vector of power generated at the nodes is denoted by $\mathbf{P}^G$ ($\mathbf{P}^G \in R^N$), where $P_i^G = 0$, if $i > L$, and let $\mathbf{P}^F$ be the vector of consumption at the nodes. If a quantity's notation also involves an $*$, it will refer to the operating point.

A natural way to formulate the relation between the power flow $\mathbf{T}$ and the $\mathbf{P}^G$ vector is the application of *Kirchhoff's nodal law* for power flow (Sect. A.4). Since the power flow $\mathbf{T}$ is with regard to the orientation of the branches and there is a power loss in the branches, a correction term must be introduced in *Kirchhoff's law*. The method will simply be as follows. If the actual flow is in accordance with the orientation, the power consumption at the endpoint will be decreased by the amount of the loss; otherwise, it will be increased by the same amount.

Let $\mathbf{P}^K \in R^N$ be the following vector (on the notation see A.4):

$$P_l^K = - \sum_{\substack{(i,l)\,\text{branch} \\ \text{sgn}\,T_{il} \geq 0}} P_{il}^v + \sum_{\substack{(i,l)\,\text{branch} \\ \text{sgn}\,T_{il} < 0}} P_{il}^v, \qquad l = 1, \ldots, N. \qquad (3.7.19)$$

Therefore, *Kirchhoff's nodal law* at operating point i will be

$$\mathbf{P}^{*G} = \mathbf{P}^F + \mathbf{P}^{*K} + \mathbf{A}^T \mathbf{T}^*. \qquad (3.7.20)$$

If we know the voltage distribution at the operating point, $\mathbf{P}^{*K}$ can be determined using (A.4.22) and (A.4.30) on the basis of (3.7.19). Having fixed the losses at the operating point, the following approximation can be applied in a small neighborhood of the operating point:

$$\mathbf{P}^G \sim \mathbf{P}^F + \mathbf{P}^{*K} + \mathbf{A}^T \mathbf{T}. \qquad (3.7.21)$$

By substituting here the approximation of the power flow as specified in (3.7.18), the following relation results:

$$\mathbf{P}^G - \mathbf{P}^F - \mathbf{P}^{*K} \sim \mathbf{A}^T \mathbf{B} \mathbf{V} \mathbf{w}. \qquad (3.7.22)$$

Here the matrix $\mathbf{A}^T \mathbf{B} \mathbf{V}$ is of size $N \times N$ and its rank is $N - 1$ (Theorem A.1.8).

If the row and column of the matrix $\mathbf{A}^T \mathbf{B} \mathbf{V}$ corresponding to the reference node are deleted, the resulting matrix is nonsingular (Theorem A.1.8). Then it is inverted, and the inverse matrix is appended with a first row and column containing 0 elements only. Let the resulting matrix be denoted by $\mathbf{Z}^N$. The phase angle of the potential at the reference node is 0; therefore, $w_1 = 0$ holds. Thus, utilizing (3.7.22), the approximation formula

$$\mathbf{w} \sim \mathbf{Z}^N (\mathbf{P}^G - \mathbf{P}^F - \mathbf{P}^{*K}) \tag{3.7.23}$$

arises, as promised in the title of this Section.

3.8 Linearizing the Network Constraints

On the basis of the discussions in Sects. 3.6 and 3.7, the opportunity to build a simplified linear model is given where in the constraints concerning the transmission network, apart from the production level and mode of operation variables, only the real parts of the complex voltages of the nodes with controllable sources of reactive power occur as variables. However, to achieve this, appropriate linear approximations to the functions appearing in the constraints (2.3.22) and (2.3.24) of the general model must be specified. Furthermore, a way of handling the system of network constraints (2.3.20) must also be specified.

First, the system of branch-load constraints (2.3.22) will be discussed. Their linearization is quite simple on the basis of (3.7.18) and (3.7.23):

$$\mathbf{T} \sim \mathbf{BVZ}^N (\mathbf{P}^G - \mathbf{P}^F - \mathbf{P}^{*K}). \tag{3.8.24}$$

Here on the right-hand side $\mathbf{P}^G$ is the only variable. Later on we will show that it can be expressed as a linear function of the production-level and mode-of-operation variables in the simplified model (Sect. 4.1.3).

In connection with the branch-load constraints, one more thing will be neglected: the apparent power $|S_{ik}|$ in constraints (2.3.22) will be replaced by the absolute value of the active power as corrected by the losses, while appropriately corrected T_{ik}^{max} bounds will be used.

Next, the system of constraints (2.3.24) concerning the reactive power sources will be considered. Applying the notations of Sect. 3.6 and disregarding the quadratic terms of the Taylor expansion of $\mathbf{g}^M (\mathbf{v}^M, \mathbf{v}^F, \mathbf{w}^*)$ around $(\mathbf{v}^{*M}, \mathbf{v}^{*F})$, we obtain

$$\mathbf{g}^M (\mathbf{v}^M, \mathbf{v}^F, \mathbf{w}^*) \sim \mathbf{g}^M (\mathbf{v}^{*M}, \mathbf{v}^{*F}, \mathbf{w}^*) + \mathbf{Y}_1 (\mathbf{v}^M - \mathbf{v}^{*M}) + \mathbf{Y}_2 (\mathbf{v}^F - \mathbf{v}^{*F}). \tag{3.8.25}$$

Substituting on the basis of (3.6.15) results in

$$\mathbf{g}^M (\mathbf{v}^M, \mathbf{v}^F, \mathbf{w}^*) \sim \mathbf{g}^M (\mathbf{v}^{*M}, \mathbf{v}^{*F}, \mathbf{w}^*) + (\mathbf{Y}_1 - \mathbf{Y}_2 \mathbf{Y}_4^{-1} \mathbf{Y}_3)(\mathbf{v}^M - \mathbf{v}^{*M}). \tag{3.8.26}$$

Finally, some remarks concerning the system of network constraints (2.3.20) follow. This consists of quadratic constraints in the form of equalities that must be satisfied by the network voltage distribution. They are implicitly included in the model using the constraints resulting from the aforementioned linearizations within a range determined by the still permitted error value in the process of linearization. The model for optimal daily scheduling, including all network constraints formulated as inequalities and with respect to the real part of the voltage of the nodes connected to controllable sources of reactive power as variables, is solved first. Then, supplementing the optimal solution by the imaginary parts of the operating-point voltages at the nodes connected to controllable sources of reactive power and based on the optimal generated powers, the load-flow problem is solved (Appendices 3 and 4). If the lower and upper bounds on the voltages at the nodes are sufficiently close to the operating point values, i.e., if the range is sufficiently narrow, then the voltage and power distribution obtained in this way will in the physical sense still satisfy the voltage, branch-load, and reactive power injection constraints. This strategy turned out to work well in practice.

3.9   Linearizing the Network Loss Function

The network loss function $C^v(\mathbf{v}, \mathbf{w})$, (2.2.10), is a convex quadratic function of the variables $(\mathbf{v}, \mathbf{w})$. Its linearization around the operating point simply means that the quadratic terms are neglected in the Taylor expansion around the point $(\mathbf{v}^*, \mathbf{w}^*)$.

Utilizing the notations of Sect. 3.6 and indicating the values corresponding to the operating point by *, we obtain the following relation:

$$P^v(\mathbf{v}, \mathbf{w}) \sim P^v(\mathbf{v}^*, \mathbf{w}^*) + \nabla^T_{\mathbf{v}^M} P^v(\mathbf{v}^*, \mathbf{w}^*)(\mathbf{v}^M - \mathbf{v}^{*M})$$

$$- \nabla^T_{\mathbf{v}^F} P^v(\mathbf{v}^*, \mathbf{w}^*)(\mathbf{v}^F - \mathbf{v}^{*F}) + \nabla^T_{\mathbf{w}} P^v(\mathbf{v}^*, \mathbf{w}^*)(\mathbf{w} - \mathbf{w}^*). \qquad (3.9.27)$$

Practical experience has shown that the third term on the right-hand side can be neglected. Let us introduce the notations $\mathbf{d}^v = \nabla_{\mathbf{v}^M} P^v(\mathbf{v}^*, \mathbf{w}^*)$, $\mathbf{b}^v = \nabla_{\mathbf{w}} P^v(\mathbf{w}^*, \mathbf{w}^*)$. The following formula can be obtained directly from Eq. (A.4.30) by computing the derivatives

$$d_i^v = \sum_{\substack{k \\ (i,k)\,\mathrm{branch}}} R_{ik} \left[2G_{ik} I_{ik}^{*P} - \left(B_{ik} - \frac{1}{2}\omega C_{ik} \right) I_{ik}^{*Q} + B_{ik} I_{ki}^{*Q} \right], \qquad i = 1,\ldots,L.$$

$$(3.9.28)$$

The components of $\mathbf{b}^v$ can be obtained in a similar way:

$$b_i^v = \sum_{\substack{k \\ (i,k)\,\mathrm{branch}}} R_{ik} \left[2B_{ik} I_{ki}^{*P} + G_{ik} I_{ik}^{*Q} - G_{ik} I_{ik}^{*Q} \right], \qquad i = 1,\ldots,N.$$

On the basis of relation (3.7.23), (3.9.27) can be formulated as follows:

$$P^v(\mathbf{v}, \mathbf{w}) \sim P^v(\mathbf{v}^*, \mathbf{w}^*) + \sum_{i=1}^{L} d_i^v v_i - \sum_{i=1}^{L} d_i^v v_i^* + \mathbf{b}^{vT} \mathbf{Z}^N (\mathbf{P}^G - \mathbf{P}^F - \mathbf{P}^{*K}) - \mathbf{b}^{vT} \mathbf{w}^*.$$

$$(3.9.29)$$

In the model constructed in the following section, it will turn out that $\mathbf{P}^G$ can be expressed as a linear function of the power and mode-of-operation variables, as mentioned previously.

3.10   Voltage Check Periods

Those periods where, in addition to caloric constraints, network constraints are also prescribed will be called *voltage check periods*. The generation of a system of network constraints and their inclusion in the optimization is coupled with a relatively higher computational time, and for this reason, in the simplified model, network constraints are only prescribed for three periods. These are the extreme load periods: the evening peak load period, the nightly minimum load period, and the morning peak load period [see Eq. (3.3.1)]. At the time of a peak load, the consumption of reactive power is high and the voltages are lower, while during the nightly minimum load period both the production of reactive power and the voltages are high. Based on general hands-on experience, which can also be derived from the strategy related to start and stop actions in power plant blocks discussed in Sect. 3.1, if network constraints are prescribed in the extreme peak load periods, they will be satisfied in the intermediate (normal) periods as well.

3.11   Selection of Network Constraints

The voltage limit constraints (2.3.21) refer to all nodes. They represent bounds on the absolute value of the complex voltages at the nodes. These will be modified by prescribing bounds only on the real parts of voltages. The reason for this is that, according to (3.7.23), restriction of the active power entails a restriction of the imaginary part of the voltages as well. On the other hand, practical experience shows that fluctuations in the imaginary part are small in the neighborhood of operating points due to the strict bounds prescribed for the other electric quantities.

In the case of bounds prescribed for the real parts of the voltages at the nodes, only a few consumer nodes are critical from the point of view of voltage limit constraints.

The operating points K_v of the most violated constraints are selected from among the constraints at the consumer nodes, or, alternatively, if those constraints are satisfied, then points are selected where the values at the operating points are closest to the limit. In practice, $K_v = 30$ proved to be sufficient.

With respect to the branch-load constraints (2.3.22), overload may occur on only a small fraction of all of the transmission lines. On the basis of the power flow at the operating point K_A, constraints are selected in a manner analogous to that used for the voltage limit constraints, and only these are prescribed in the simplified model. In practice, $K_A = 30$ proved to be sufficient here as well.

Chapter 4
The Model Obtained by Taking into Account the Simplifying Assumptions

On the basis of the discussion in Chap. 3, the general model of the scheduling problem can be simplified.

The simplified model is a large-scale mixed-variable mathematical programming problem with a linear objective function and linear constraints, with the coefficient matrix having a special structure. It is suitable for numerical solutions.

In the process of formulating the model, some changes in the notations take place, apart from the already mentioned simplifications: in the simplified model, the mode-of-operation variables are defined differently from how they were defined in the general model to decrease the number of 0–1 variables.

The notations used in the description of the model here are the same as those used in Sect. 2.1.

4.1 Simplified Model

4.1.1 Mode-of-Operation Variables

The x_{ij}^t, $i = 1, 2, \ldots, K$; $j = 1, 2, \ldots, M(i) - 1$; $t = 1, 2, \ldots, 27$ variables will be defined in the following way [K is the number of power plants, and $M(i)$ is the number of modes of operation applicable at power plant i, i.e., as in the case of the general model]:

$$
x_{ij}^t = \begin{cases}
0 & \text{if at the } i\text{th power plant in the } t\text{th period the } j\text{th mode of operation or} \\
& \text{a preceding mode of operation is active;} \\
1 & \text{if at the } i\text{th power plant in the } t\text{th period a mode of operation} \\
& \text{following the } j\text{th mode of operation is active.}
\end{cases}
$$

$$(4.1.1)$$

A. Prékopa et al., *Scheduling of Power Generation*, Springer Series in Operations Research and Financial Engineering, DOI 10.1007/978-3-319-07815-1_4,
© Springer International Publishing Switzerland 2014

It is obvious that if the interpretation of the symbols $x^t_{i,M(i)}$, $i = 1, 2, \ldots, K$, $t = 1, 2, \ldots, 27$, is also defined in the foregoing way, their value can only be 0. Therefore, $x^t_{i,M(i)} = 0$, $i = 1, 2, \ldots, K$, $t = 1, 2, \ldots, 27$, and they are not variables of the model. The number of modes of operation in the case of individual power plants is only $M(i) - 1$.

Despite this, the symbols $x^t_{i,M(i)}$ defined in the preceding sense occur in the model description. Similarly, the symbols $x^t_{i,0}$, $i = 1, 2, \ldots, K$, $t = 1, 2, \ldots, 27$, will also be used in the model formulation. Their value is 1, in accordance with the definition given previously.

Given this definition, the value of the difference of $x^t_{i,j-1} - x^t_{ij}$ is 1 if and only if in the tth period at the ith power plant the jth mode of operation is active $[i = 1, 2, \ldots, K, \ j = 1, 2, \ldots, M(i), \ t = 1, 2, \ldots, 27]$. It also follows from the definition of x^t_{ij} that the components corresponding to a given power plant, i.e., the $x^t_{i1}, x^t_{i2}, \ldots, x^t_{i,M(i)-1}$ variables, take a value of 1 for a first section of consecutive indices, followed by a second section where the value is 0 throughout. Concerning the pair 1, 0 standing side by side, the position of 0 indicates which mode of operation is operating at the given power plant in a particular period.

In the foregoing groups of variables, corresponding to different periods, the position of the 1, 0 pair can vary. A move to the right means that a stop action took place at the power plant in the time interval between the two periods. If it moves to the left, then a startup action has occurred. If the position of the pair of 1, 0 is unchanged, the mode of operation has not changed in the time interval between the periods (Sect. 3.2).

The $\mathbf{x}^t$ vector of the mode-of-operation variables corresponding to the tth period is provided by concatenating the x^t_{ij}, $i = 1, 2, \ldots, K$, $j = 1, 2, \ldots, M(i) - 1$ components in the order of the power plants and for a fixed power plant in the order of the modes of operation. Its dimension is $\sum_{i=1}^{K}(M(i) - 1)$.

Due to the assumption in Sect. 3.1, there is no changeover in the stagnation phases. Therefore, it is quite sufficient to regard the mode-of-operation variables of the first period of the stagnation phase as the model variables. For this reason, $\mathbf{x}^t$ only denotes variables in the case of $t = 1, 2, \ldots, t_0, t_0 + 4, \ldots t_1, t_1 + 4, \ldots, 27$, where t_0 and t_1 are the serial numbers of the first periods of the phases of stagnation. The $\mathbf{x}$ variable without a superscript indicates the concatenation of the $\mathbf{x}^t$ variables. The dimension of $\mathbf{x}$ is

$$21 \sum_{i=1}^{K}(M(i) - 1).$$

In the formal description of the model the symbol x^t_{ij} occurs also in the case of $t = t_0 + 1$, $t_0 + 2$, $t_0 + 3$ and $t_1 + 1$, $t_1 + 2$, $t_1 + 3$ as well. These are to be understood as $x^{t_0}_{ij}$ respectively $x^{t_1}_{ij}$. In the model it is necessary to have information about the mode of operations in the last period preceding the current planning stage. The relevant information can be provided by supplying the value of the mode-of-operation variables of this last period. Let us denote the corresponding vector of

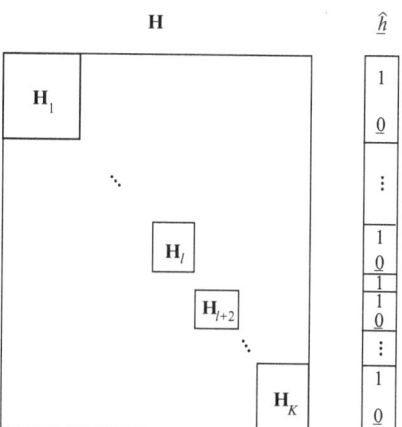

Fig. 4.1 Structure of matrix and constant term of linear transformation connecting mode-of-operation variables of general and simplified model

the modes of operation by $\mathbf{x}^0$. In the model this is a $\sum_{i=1}^{K}(M(i)-1)$-dimensional, constant vector of 0, 1 values.

Note that if the modes of operation and their order in the general model are the same as the modes of operation and their order in this model, then for the mode-of-operation variables of the two models the following relation holds:

$$\mathbf{y}^t = \mathbf{H}\mathbf{x}^t + \hat{\mathbf{h}}. \qquad (4.1.2)$$

To avoid notation that is too complicated, instead of providing $\mathbf{H}$ and $\hat{\mathbf{h}}$ in an explicit form, they are visualized in Fig. 4.1.

In the figure, the H_l block is of size $M(l) \times (M(l)-1)$ and $(\mathbf{H}_l)_{kk} = -1$, $k = 1, 2, \ldots, M(l)-1$, $(\mathbf{H}_l)_{k+1,k} = 1$, $k = 1, 2, \ldots, M(l)-1$. The remaining elements of the matrix are 0. The rows corresponding to those power plants where there is only one mode of operation contain only 0s. In the figure there is a row of this type between the $\mathbf{H}_l$ and $\mathbf{H}_{l+2}$ blocks. The corresponding component in vector $\hat{\mathbf{h}}$ is 1.

4.1.2 Production-Level Variables

To specify the production level of the modes of operation to be applied in the individual periods at power plants, $r(i, j)$ variables are attached to each mode of operation in each period, where $r(i, j)$ is the number of line segments in the piecewise-linear approximation to the cost function corresponding to the mode of operation.

Let p_{ijl}^t, $l = 1, 2, \ldots, r(i, j)$, $t = 1, 2, \ldots, 27$, denote the variable attached to the jth mode of operation at the ith power plant.

Let $p_{ijl}^t = 0, l = 1, 2, \ldots, r(i, j)$, if in the tth period at the ith power plant the active mode of operation is not the jth mode of operation. Otherwise, in the case where the jth mode of operation of the ith power plant is active on production level P in the tth period, let $p_{ijl}^t \geq 0, l = 1, 2, \ldots, r(i, j)$, and, provided that

$$P_{ijk}^{\min} \leq P \leq P_{ijk}^{\max} \tag{4.1.3}$$

holds, let

$$p_{ijl}^t = P_{ijl}^{\max} - P_{ijl}^{\min}, \qquad l = 1, 2, \ldots, k - 1, \tag{4.1.4}$$

$$p_{ijk}^t = P - P_{ijk}^{\min}, \tag{4.1.5}$$

$$p_{ijl}^t = 0, \qquad l = k + 1, \ldots, r(i, j). \tag{4.1.6}$$

From the definition it follows that $p_{ijk}^t > 0$ can only hold for some index k if (4.1.4) is fulfilled. The definition also implies that the conditions

$$0 \leq p_{ijl}^t \leq (P_{ijl}^{\max} - P_{ijl}^{\min})(x_{ij-1}^t - x_{ij}^t), \tag{4.1.7}$$

$$i = 1, 2, \ldots, K, \quad j = 1, 2, \ldots, M(i), \quad l = 1, 2, \ldots, r(i, j)$$

$$t = 1, 2, \ldots, 27$$

must also hold. In these relations, the factor $(x_{ij-1}^t - x_{ij}^t)$ on the right-hand side ensures that $p_{ijl}^t > 0$ can only hold in the case where the jth mode of operation is active at the ith power plant.

Using the previously defined production-level variables, the production level of the ith power plant in the tth period can be specified by the following sum:

$$P = P_{ij}^{\min}(x_{ij-1}^t - x_{ij}^t) + \sum_{l=1}^{r(i,j)} p_{ijl}^t. \tag{4.1.8}$$

Adding up these relations for all modes of operation at the ith power plant, the production level of the power plant in the tth period results in

$$\sum_{j=1}^{M(i)} \left[P_{ij}^{\min}(x_{i,j-1}^t - x_{ij}^t) + \sum_{l=1}^{r(i,j)} p_{ijl}^t \right]. \tag{4.1.9}$$

For a fixed t, let $\mathbf{p}^t$ denote a vector that is formed by the p_{ijl}^t components by enlisting them according to the ordering of the power plants and their modes of operation.

This $\mathbf{p}^t$ will be the vector of production-level variables in the model corresponding to the tth period. Its dimension is

$$\sum_{i=1}^{K}\sum_{j=1}^{M(i)} r(i,j).$$

The vector **p** without a superscript will be the vector of production-level variables in the model. This vector is formed by concatenating the vectors $\mathbf{p}^t$, $t = 1, 2, \dots, 27$.

4.1.3   Voltage Variables

In the voltage check periods, the real parts of voltages at the nodes corresponding to controllable sources of reactive power are also variables of the model, in addition to production-level and mode-of-operation variables. Let **u** denote the L-dimensional vector formed by the preceding variables.

It was assumed in Chap. 3 that the $\mathbf{P}^G$ vector of the active powers generated at the nodes could be expressed as a linear function of the variables **p**, **x** of the model. Indeed, let $\mathbf{D}^*$ be the following matrix of size $N \times N_v$ [where N_v denotes the dimension of $(\mathbf{p}^T, \mathbf{x}^T)$]: its nonzero elements can only be found in those rows corresponding to nodes that a power plant is connected to. In these rows, in the columns of power type variables corresponding to the connected power plants, there is a 1, while the corresponding $P_{i,j+1}^{\min} - P_{ij}^{\min}$ quantities can be found in the columns of mode-of-operation type variables. It is easy to see that

$$\mathbf{P}^G = \mathbf{D}^* \begin{pmatrix} \mathbf{p} \\ \mathbf{x} \end{pmatrix} + \mathbf{P}^{\min} \tag{4.1.10}$$

holds, where $P_k^{\min}$ is the sum of the minimum production levels $P_{i1}^{\min}$ of the modes of operation belonging to power plants connected to the kth node, and $P_k^{\min} = 0$ if there is no power plant connected to the kth node.

4.2   Objective Function

Considering the simplifying assumptions explained in Chap. 3 and utilizing the variables of the simplified model, the components of the nonlinear objective function discussed in Sect. 2.2 are altered in the following way.

4.2.1 Production Costs of Power Plant Blocks

The cost of production of the jth mode of operation of the ith power plant in the tth period, corresponding to the production level determined by the production-level variables p_{ijl}^t, $l = 1, 2, \ldots, r(i, j)$ according to (4.1.8), will be

$$a_t \left\{ K_{ij} (x_{i,j-1}^t - x_{ij}^t) + \sum_{l=1}^{r(i,j)} c_{ijl} \cdot p_{ijl}^t \right\}, \tag{4.2.11}$$

where the piecewise-linear approximations to the production cost functions $f_{ij}(\mathbf{P})$, $i = 1, 2, \ldots, K$, $j = 1, 2, \ldots, M(i)$, have been utilized.

In the entire planning stage the partial costs of the operation of the power plant blocks is

$$\sum_{t=1}^{27} a_t \sum_{i=1}^{K} \sum_{j=1}^{M(i)} \left\{ K_{ij} (x_{i,j-1}^t - x_{ij}^t) + \sum_{l=1}^{r(i,j)} c_{ijl} p_{ijl}^t \right\} \tag{4.2.12}$$

which is a linear function of the model variables.

4.2.2 Partial Costs Due to Standstill and Restart

In accordance with Sect. 3.4, the approximate value of the costs of standstill of power plant units respectively modes of operation can be obtained by adding up the d_{ij}^t quantities.

According to the definition of mode-of-operation variables and the assumption on the specification of the modes of operation, if $x_{ij}^t = 1$ holds, then at the ith power plant in the tth period a mode of operation following the jth mode of operation is active. However, this implies that at least one unit belonging to mode of operation j is in standstill.

Therefore, the addition of the d_{ij}^t quantities over the standstill time corresponds to the addition of the $d_{ij}^t x_{ij}^t$ products for the entire planning stage (excluding the second, third, and fourth periods of the stagnation phases). Consequently, the cost of standstill is

$$\sum_{t=1}^{27} \sum_{i=1}^{K} \sum_{j=1}^{M(i)} d_{ij}^t x_{ij}^t, \tag{4.2.13}$$

where in the summation $t \neq t_0 + 1, t_0 + 2, t_0 + 3$ and $t \neq t_1 + 1, t_1 + 2, t_1 + 3$ hold.

This sum, however, contains the costs corresponding to units already in standstill at the end of the preceding planning stage, but these should be neglected according to the simplifying assumptions. Thus, the value of (4.2.13) must be modified. For this let us change the definition of the d_{ij}^t coefficients. Let $d_{ij}^t = 0$ for every t value belonging to the first stop phase, to the first start phase, to the first period of the first stagnation phase and every mode of operation i, j [$i = 1, 2, \ldots, K$, $j = 1, 2, \ldots, M(i)$] for which $x_{ij}^0 = 1$ holds ($x_{ij}^0 = 1$ means that in the ith power plant at the end of the preceding planning stage a mode of operation following the jth mode of operation is active; therefore, a power plant unit belonging to the jth mode of operation is in standstill).

If calculating the standstill costs using the altered values of the d_{ij}^t coefficients in (4.2.13), there is no cost of standstill in the first part of the planning stage corresponding to units of types ① and ② (using the notations of Fig. 3.2) that are shut off.

In the second part of the day, the costs of standstill of units of types ② and ④ are still calculated, despite the fact that this should be neglected according to the simplifying assumptions. A characteristic feature of stops and restarts of types ② and ④ is that the shutoff unit is in standstill already in the last period of the first startup phase. Let t^* denote the serial number of the aforementioned period.

If $x_{i_0 j_0}^{t^*} = 1$ holds, then a shutoff unit of types ② and ④ belongs to the j_0th mode of operation in the i_0th power plant. In the case of the addition of the $d_{i_0 j_0}^t$ values for the entire planning stage, the approximate value $s_{i_0 j_0}(4 + 2l_0)$ occurs in the sum for the second part of the day, provided that l_0 is the number of periods of the second stop phase. Let us decrease the cost of standstill (4.2.13) by the sum

$$\sum_{i=1}^{K} \sum_{j=1}^{M(i)} s_{ij}(4 + 2l_0)x_{ij}^{t^*}. \tag{4.2.14}$$

If for the case of $t = t^*$ the d_{ij}^t coefficients are altered in such a way that their value decreases by $s_{ij}(4 + 2l_0)$, [$i = 1, 2, \ldots, K$, $j = 1, 2, \ldots, M(i)$], then (4.2.13) is the approximation of the cost of standstill in accordance with our simplifying assumptions.

4.2.3 Costs of Transmission Losses

Let the considered voltage check period be the tth one, and where no misunderstanding may occur, superscript t is neglected concerning the variables. Let $F(\mathbf{u}, \mathbf{p}, \mathbf{x})$ denote the value of transmission losses, in the local currency (Fts), for the period. On the basis of (3.9.29), the partial costs in the objective function due to the losses are as follows:

$$f(\mathbf{u},\mathbf{p},\mathbf{x}) = \gamma a_t \left[\mathbf{d}^T \mathbf{u} + \mathbf{b}^T \mathbf{Z}^N \left(\mathbf{D}^* \begin{pmatrix} \mathbf{p} \\ \mathbf{x} \end{pmatrix} + \mathbf{P}^{\min} - \mathbf{P}^F - \mathbf{P}^{*K} \right) \right], \qquad (4.2.15)$$

where γ is the cost of the loss of 1 MWh in Fts and a_t is the length of the tth period in hours. The total cost of transmission losses is the sum of the costs of losses of the individual periods.

4.3   Model Constraints

The constraints of the simplified model can also be arranged in groups by whether they describe connections in a repeated way for the periods only among variables of the individual periods or prescribe relations among the variables of several periods. Among the constraints, there are some that are repeated for the periods, and they must be satisfied explicitly only in the voltage check periods.

4.3.1   System of Constraints of a Normal Period

The satisfaction of the following constraints is required in every period. *Normal period* refers to those periods where, except for these constraints, no further constraints are needed to describe the relationships among the variables of these periods.

By the definition of production-level variables, the fulfillment of the

$$0 \le p_{ijl}^t \le (P_{ijl}^{\max} - P_{ijl}^{\min})(x_{ij-1}^t - x_{ij}^t) \qquad (4.3.16)$$

$$i = 1,2,\dots,K, \ j = 1,2,\dots,M(i), \ l = 1,2,\dots,r(i,j)$$

variable coupling constraints must be required for each of the periods.

There is no need to include a constraint in the model to ensure that $p_{ijk}^t > 0$ holds only in the case of $p_{ijl}^t = P_{ijl}^{\max} - P_{ijl}^{\min}$, $l = 1,2,\dots,k-1$. Due to the property (3.3.2) of the approximate functions of the production cost functions $f_{ij}(P)$, this is automatically satisfied for the optimal solutions with minimal costs of the mixed-variable problem corresponding to the model.

The *supply constraint* can be specified in the following way by using the variables of the simplified model:

$$\sum_{i=1}^{K} \sum_{j=1}^{M(i)} (x_{ij-1}^t - x_{ij}^t) P_{ij}^{\min} + \sum_{l=1}^{r(i,j)} p_{ijl}^t = P^{t\ \mathrm{dem}} + P^{t\ \mathrm{loss}} + P^{t\ \mathrm{self}}. \qquad (4.3.17)$$

Though this fact was not explained in the discussion of the simplifying assumptions, the functions of self-consumption of the power plants (Sect. 2.3.1) are assumed in this model to be independent of the modes of operation and are approximated by linear functions of power plant production. Therefore,

$$P^{t \text{ self}} = \sum_{i=1}^{K} P_i^{\text{ self}} \left(\sum_{j=1}^{M(i)} (x_{ij-1}^t - x_{ij}^t) P_{ij}^{\text{min}} + \sum_{l=1}^{r(i,j)} p_{ijl}^t \right) \tag{4.3.18}$$

holds, where $P_i^{\text{ self}}(P)$ is the linear function of self-consumption at power plant i.

In the simplified model there is no need for the SOS constraints of the general model. It is, however, necessary to require the fulfillment of the relation

$$x_{ij-1}^t - x_{ij}^t \geq 0, \qquad j = 1, 2, \ldots, M(i) - 1, \tag{4.3.19}$$

for $i = 1, 2, \ldots, K$, due to the definition of the modes of operation. These are referred to as the *constraints on the mode-of-operation variables*.

4.3.2 System of Constraints of Voltage Check Periods

Besides the constraints required in the normal periods, additional constraints are to be prescribed in the voltage check periods.

Relying on the contents of Sects. 2.3.1 and 3.5–3.8, the system of constraints on the transmission network in the simplified model is specified here. Because a fixed period is being considered here, superscript t will be omitted. Quantities referring to the operating point are indicated by an $*$ in a superscript position.

4.3.3 Voltage Limit Constraints

The V_i^{min}, V_i^{max}, $i = 1, \ldots, N$ bounds referring to the absolute value of the voltages at the nodes are given for each node. The bounds corresponding to the real part are calculated as follows:

$$u_i^{\text{min}} = [(V_i^{\text{min}})^2 - (w_i^*)^2]^{\frac{1}{2}}, \quad u_i^{\text{max}} = [(V_i^{\text{max}})^2 - (w_i^*)^2]^{\frac{1}{2}}, \quad i = 1, \ldots, N. \tag{4.3.20}$$

The voltage limit constraints corresponding to the nodes connected to controllable sources of reactive power are individual lower upper bounds:

$$u_i^{\text{min}} \leq u_i \leq u_i^{\text{max}}, \qquad i = 1, \ldots, L. \tag{4.3.21}$$

With respect to the consumer nodes, relation (3.6.15) can be applied, leading to the following system of constraints:

$$\mathbf{u}^{\min F} \leq \mathbf{v}^{*F} - \mathbf{Y}_4^{-1}\mathbf{Y}_3(\mathbf{u} - \mathbf{v}^{*M}) \leq \mathbf{u}^{\max F}, \qquad (4.3.22)$$

where the notations of (3.5.6) were used.

4.3.4 Branch-Load Constraints

Let $\mathbf{T}^{\max}$ denote the M-dimensional vector of the $T_{ik}^{\max}$ quantities in (2.3.22) that were corrected in accordance with Sect. 3.8. On the basis of relations (3.7.23) and (3.7.18),

$$\mathbf{T} \sim \mathbf{BVZ}^N(\mathbf{P}^G - \mathbf{P}^F - \mathbf{P}^{*K}) \qquad (4.3.23)$$

results. Taking into account the relation concerning $\mathbf{P}^G$ obtained in Sect. 4.1.3, the system of branch-load constraints is as follows:

$$-\mathbf{T}^{\max} \leq \mathbf{BVZ}^N\left[\mathbf{D}^*\begin{pmatrix}\mathbf{p}\\\mathbf{x}\end{pmatrix} + \mathbf{P}^{\min} - \mathbf{P}^F - \mathbf{P}^{*K}\right] \leq \mathbf{T}^{\max}. \qquad (4.3.24)$$

Remark: Note that, in accordance with the discussion in Sect. 3.11, the model includes K_v respectively K_A voltage limit constraints corresponding to the consumer nodes respectively branch-load constraints.

4.3.5 Reactive Power Source Constraints

This system of constraints can be formulated easily on the basis of (3.8.26) and (2.3.24):

$$\mathbf{Q}^{\min}(\mathbf{x}) \leq \mathbf{g}^M(\mathbf{v}^*, \mathbf{w}^*) + (\mathbf{Y}_1 - \mathbf{Y}_2\mathbf{Y}_4^{-1}\mathbf{Y}_3)(\mathbf{u} - \mathbf{v}^{*M}) \leq \mathbf{Q}^{\max}(\mathbf{x}). \qquad (4.3.25)$$

Here the notations of Sect. 3.8 are used and the $\mathbf{Q}^{\min}(\mathbf{x})$ and $\mathbf{Q}^{\max}(\mathbf{x})$ vectors, consisting of the components $Q_i^{\min}(\mathbf{x})$ and $Q_i^{\max}(\mathbf{x})$, $i = 1,\ldots,L$, are introduced. The functions on both the left- and right-hand sides of (4.3.25) must be specified as well. The considerations concerning the system of constraints (2.3.24) in the general model are valid here, too. Therefore, to specify the dependence of the lower and upper bounds on $\mathbf{x}$, it is quite sufficient to consider the linear relation (4.1.2) between the $\mathbf{y}$ vector of the modes of operation of the general model and the $\mathbf{x}$ vector of the modes of operation.

On this basis, the $Q_i^{\min}(\mathbf{x})$ and $Q_i^{\max}(\mathbf{x})$ functions take the following form, with notations as in (2.3.24):

$$Q_i^{\min}(\mathbf{x}) = (\mathbf{H}\mathbf{x} + \hat{\mathbf{h}})^T \hat{\mathbf{Q}}_i^{\min}, \quad Q_i^{\max}(\mathbf{x}) = (\mathbf{H}\mathbf{x} + \hat{\mathbf{h}})^T \hat{\mathbf{Q}}_i^{\max}, \quad i \in I_M \cap I_E,$$
$$Q_i^{\min}(\mathbf{x}) = Q_i^{\min}, \quad\quad\quad\quad Q_i^{\max}(\mathbf{x}) = Q_i^{\max}, \quad\quad\quad\quad i \in I_M \setminus I_E.$$

$$(4.3.26)$$

4.3.6 Constraints Connecting the Periods

In the simplified model there is no need for constraints corresponding to the stop-and-start constraints in the general model. In this section it is sufficient to specify the mathematical formulation of the regulations concerning operations management as discussed in Sect. 3.1.

Start actions are not permitted in a stop phase. If the tth period is the last period of the start phase or belongs to a stop phase, then

$$x_{ij}^t = 1 \quad\quad \text{implies} \quad\quad x_{ij}^{t+1} = 1.$$

This Boolean condition is best described by the inequality

$$x_{ij}^t \leq x_{ij}^{t+1}. \tag{4.3.27}$$

The matrix of the resulting constraints is shown in Fig. 4.2.

The start conditions can be described in an analogous way. If t is the serial number of the first period of a stagnation phase or not the last period of a start phase, then

$$x_{ij}^t = 0 \quad\quad \text{implies} \quad\quad x_{ij}^{t+1} = 0.$$

In the form of an inequality the constraint is

$$x_{ij}^t \geq x_{ij}^{t+1}. \tag{4.3.28}$$

The matrix structure of these constraints is displayed in Fig. 4.3.

4.3.7 Fuel Constraints

Fuel constraints are also constraints connecting several periods. To describe them, it is sufficient to refer to Sect. 2.3.2 with respect to the general model and rewrite formula (2.3.29) in terms of the variables of the simplified model:

$$R_{i\,\min} \le \sum_{t=1}^{27} a_t \sum_{j=1}^{M(i)} \left\{ P_{ij}^{\min}(x_{ij-1}^t - x_{ij}^t) + \sum_{l=1}^{r(i,j)} p_{ijl}^t \right\} \le R_{i\,\max}. \qquad (4.3.29)$$

Let

$$h_i(\mathbf{p},\mathbf{x}) = \sum_{t=1}^{27} a_t \sum_{j=1}^{M(i)} \left\{ P_{ij}^{\min}(x_{ij-1}^t - x_{ij}^t) + \sum_{l=1}^{r(i,j)} p_{ijl}^t \right\} \qquad (4.3.30)$$

be the linear function of the **p** production-level variables and the **x** mode-of-operation variables. Based on this notation, the fuel constraints can be written as

$$R_{i\,\min} \le h_i(\mathbf{p},\mathbf{x}) \le R_{i\,\max}. \qquad (4.3.31)$$

4.4   Structure, Characteristics, and Size of the Simplified Model

The next step is to formulate the large-scale mixed-variable mathematical programming problem, with a linear objective function and linear constraints, corresponding to the simplified model.

The objective function to be minimized is provided by the sum of the costs of standstill in (4.2.13) decreased by (4.2.14), the production costs (4.2.12), and the costs (4.2.15) of transmission losses.

The system of constraints consists of the following items:

(4.3.16)	variable coupling constraints,
(4.3.17)	supply constraints,
(4.3.19)	constraints on mode-of-operation variables,
(4.3.22)	voltage limit constraints,
(4.3.24)	branch-load constraints,
(4.3.25)	reactive power source constraints,
(4.3.27)–(4.3.28)	stop-and-start constraints,
(4.3.30)	fuel constraints.

Figure 4.4 shows the coefficient matrix of this problem. Variables in the individual periods are in the following order: voltage variables, production-level variables, mode-of-operation variables. (Recall that only a few periods have associated voltage variables, and the mode-of-operation variables do not correspond to the second, third, and fourth periods of the stagnation phases.)

Each period is associated with a block of constraints, including variables corresponding to the given period only. In Fig. 4.4 blocks corresponding to normal periods are indicated by ①, while blocks of voltage check periods are indicated by ②.

The blocks connected by a dotted line in Fig. 4.4 comprise the constraints of the stop, stagnation, and start phases. Apart from the blocks denoted by ① and

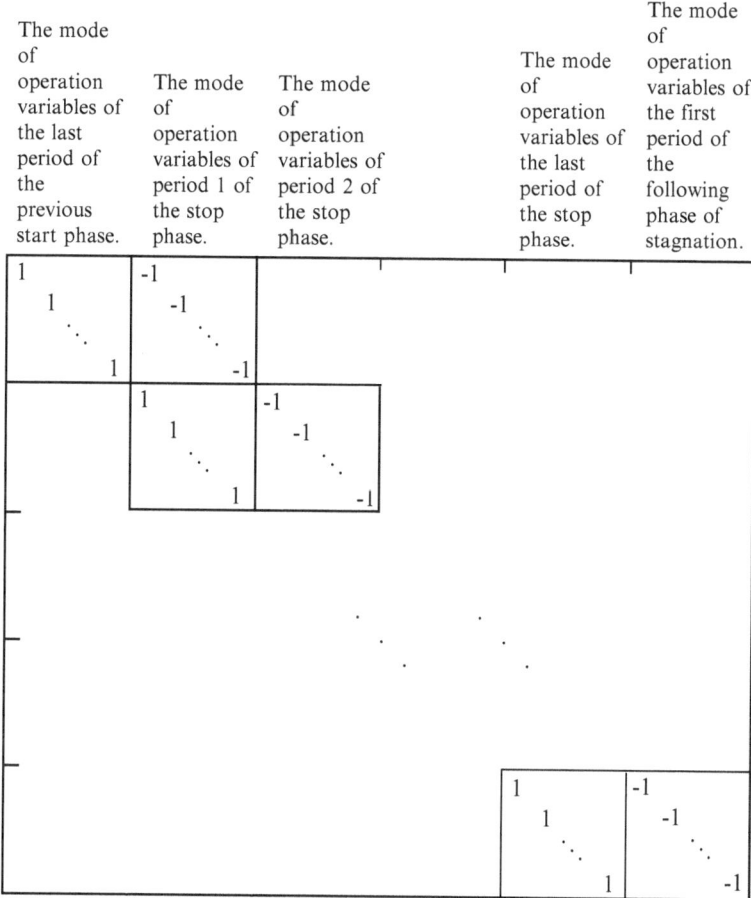

Fig. 4.2 Matrix of stop constraints

②, blocks of the stop phase (denoted by ③ and ⑥ in Fig. 4.4) include the stop conditions, while blocks of the start phase (⑤, ⑧) include the start conditions. Blocks corresponding to the stagnation phase (④, ⑦) contain a single vector of the mode-of-operation variables. The "small blocks" belonging to them and corresponding to the periods are connected to one another by this vector of mode-of-operation variables.

The same vectors of mode-of-operation variables establish connections between blocks ③ and ⑤ and between ⑥ and ⑧, respectively. Blocks ⑤ and ⑥ are connected to one another by the vector of mode-of-operation variables belonging to the last period of the first start phase.

Constraints connecting all the periods (top of Fig. 4.4) are fuel constraints.

The size of the mixed-variable linear programming problem, corresponding to the simplified model, is as follows.

The mode of operation variables of the first period of the previous phase stagnation.

The mode of operation variables of period 1 of the start phase.

The mode of operation variables of period 2 of the start phase.

The mode of operation variables of the last period of the start phase.

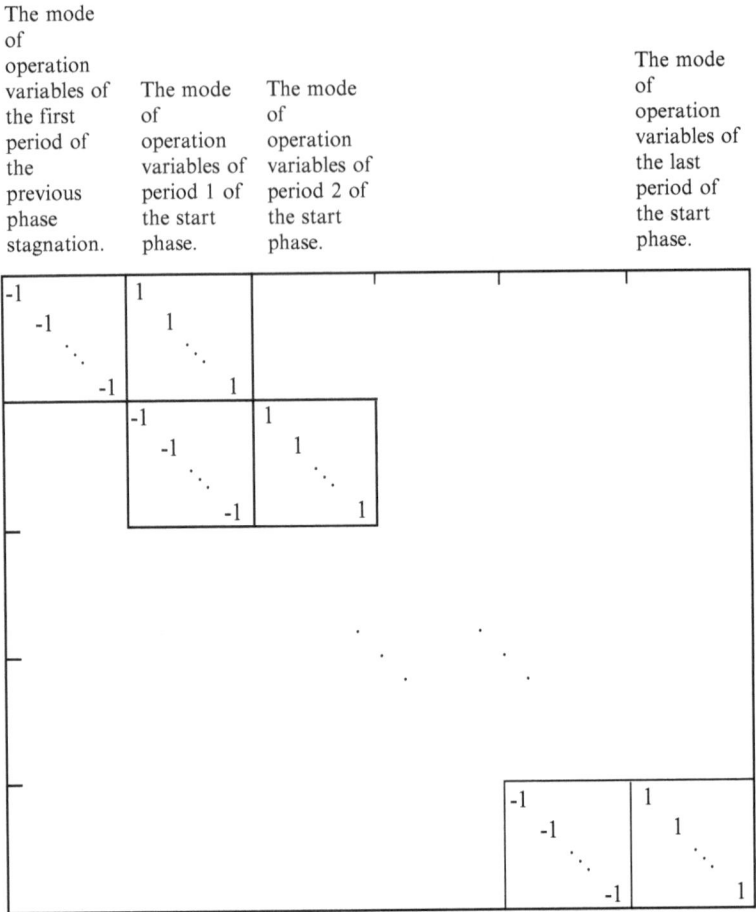

Fig. 4.3 Matrix of start constraints

4.4.1 Number of Variables

1. Number of 0–1 variables $= 21x$:
   the difference between the number of modes of operation applicable in the electric power generating system and the number of power plants;
2. Number of production-level variables $= 27x$:
   the overall sum of the number of approximate sections of the production cost functions of the modes of operation applicable in the electric power generation system;
3. Number of voltage variables $= 3x$:
   the number of nodes connected to controllable sources of reactive power in the network.

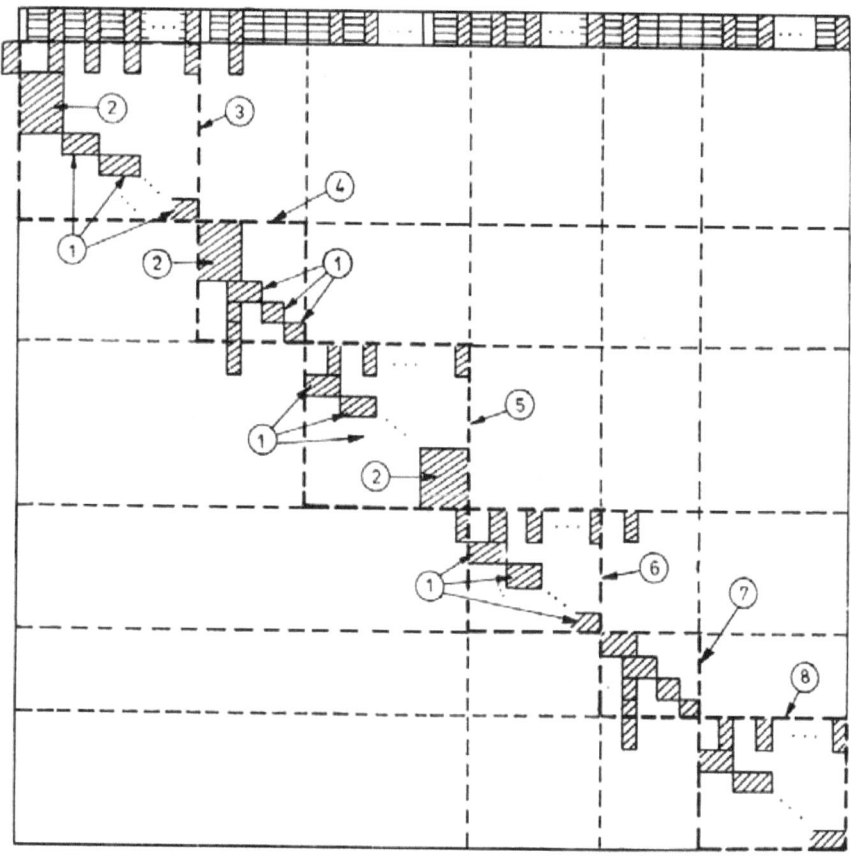

Fig. 4.4 Structure of simplified model

4.4.2 Number of Constraints

1. Periodwise, in each normal period (24 times in the problem): as many variable coupling constraints as production-level variables + one supply constraint + as many constraints as mode-of-operation variables restricted by them (the number of mode-of-operation variables is 0 in the second, third, and fourth periods of the stagnation phase);
2. In addition to the constraints of a normal period, in the case of a voltage check period $2\times$ the number of nodes connected to controllable sources of reactive power + 60 constraints appear, including individual lower and upper bounds of as many nodes connected to controllable sources of reactive power;
3. The number of stop–start constraints $= 21\times$ the number of modes of operation applicable in the power generation system;
4. At most five fuel constraints.

The solution approach to this problem will be discussed in the subsequent Section.

4.5   Summary of Notations Introduced in Chaps. 3 and 4

$s_{ij}(\tau)$	Function of standstill costs of jth mode of operation at ith power plant;
$r(i,j)$	Number of line segments in approximation of function $f_{ij}(P)$;
$P_{ijl}^{\min}, P_{ijl}^{\max}$	Levels of production belonging to endpoints of lth line segment in approximation;
c_{ijl}	Slope of lth approximating line segment;
d_{ij}^t	Part of cost of standstill of jth mode of operation at ith power plant to be taken into account in period t;
$\mathbf{x}^t, x_{ij}^t$	Mode-of-operation variable respectively its component corresponding to tth period;
$\mathbf{x}^0$	Vector of modes of operation of last period of day preceding current planning stage;
t_0, t_1	Serial number of period corresponding to beginning of first and second stagnation phases;
p_{ijl}^t	$l = 1, 2, \ldots, r(i,j)$, component of production-level variable providing production level of jth mode of operation at ith power plant in tth period;
$\mathbf{p}^t, \mathbf{p}$	Vector of production-level variables of tth period and vector of production-level variables corresponding to whole scheduling interval;
t^*	Serial number of last period of first start phase;
$h_i(\mathbf{p}, \mathbf{x})$	Linear function providing daily production of ith power plant;
L	Number of nodes connected to controllable sources of reactive power ($= N_M$);
N_F	Number of consumer nodes, $L + N_F = N$;
$\mathbf{u}$	Vector of voltage variables in fixed period; its components are formed by real parts of voltages of nodes connected to controllable sources of reactive power, $\mathbf{u} \in R^L$;
$(\mathbf{v}^*, \mathbf{w}^*)$	Vector of voltages at operating point;
$u_i^{\min}$	Lower bound on real part of voltages at nodes connected to controllable sources of reactive power;
$u_i^{\max}$	Upper bound on real part of voltages at nodes connected to controllable sources of reactive power;

B	Diagonal matrix of size $M \times M$, and its diagonal elements are the B_{ik} quantities (see Sect. A.3);
V	Matrix of size $N \times N$ needed for compact formulation of dependence of power flow on voltage, introduced after relation (3.7.17) (see its definition there);
A	Incidence matrix of network; in notation of appendix the node-edge incidence matrix of graph $(\mathcal{N}_\gamma, \mathcal{A}_\gamma^D)$;
$\mathbf{Z}^N$	Can be obtained as follows: the row and column corresponding to the reference node should be deleted in the matrix $\mathbf{A}^T \mathbf{B} \mathbf{V}$, and the resulting matrix should be appended by a **0** row and column in positions of deleted rows;
$\mathbf{Y} = \begin{pmatrix} \mathbf{Y}_1 & \mathbf{Y}_2 \\ \mathbf{Y}_3 & \mathbf{Y}_4 \end{pmatrix} \begin{matrix} \}L \\ \}N_F \end{matrix}$ $\underbrace{}_{L}\underbrace{}_{N_F}$	*Jacobi matrix* of transformation $g(\mathbf{v}, \mathbf{w}^*) : R^N \to R^N$ partitioned according to nodes connected to controllable sources of reactive power and consumer nodes;
D*	Matrix of linear transformation providing actual injection of real power at nodes by transforming production-level and mode-of-operation variables;
$\mathbf{P}^K$	Correction vector of branchwise real power losses;
$\mathbf{d}^v$	Coefficient of **u** in part of objective function representing transmission losses;
$\mathbf{b}^v$	Vector needed for construction of coefficients of **p**, **x** variables in part of objective function representing transmission losses;
γ	Cost of 1-MWh loss in local currency (Fts).

Chapter 5
Daily Scheduling

Both the generation system and the network transmitting electric power vary on a day-to-day basis; therefore, the daily optimization problems to be solved and their sizes also vary, although their structure remains essentially the same. This chapter shows how the daily scheduling problem, corresponding to the simplified model of the scheduling problem, can be set up and solved.

5.1 Generating the Mixed-Variable Problem Corresponding to Daily Data

Because of necessary maintenance work, failure, and various fitting activities, the transmission capacity of individual branches in a network may vary within a day (in the case, for example, of shunts or transformers), or they might not be available for carrying electric energy at all. The situation is similar with power plants; because of failure, maintenance, or other issues, the production capacity of power plants and the available modes of operation vary from day to day. Therefore, the numerical optimization problem instance of the daily scheduling problem must be set up (generated) on a daily basis. This means that, using adequate databases containing constant data, these data should be retrieved and modified based on the present state of the power system. The following factors must be determined for each particular day: the general situation, the available network, the actual transmission capacity of its different branches, the available modes of operation at the power plants, along with their limitations and constraints.

A separate problem is presented by the setting up of the system of constraints for daily voltage check periods. First a load-flow problem must be solved to determine the operating point and the corresponding transmission losses. Then the system of constraints corresponding to this operating point and transmission losses is computed (only some of the constraints are taken into account; see Sect. 3.11). Note that the voltage distribution is recalculated on the basis of the voltage data

A. Prékopa et al., *Scheduling of Power Generation*, Springer Series in Operations Research and Financial Engineering, DOI 10.1007/978-3-319-07815-1_5,
© Springer International Publishing Switzerland 2014

at the nodes connected to controllable sources of reactive power obtained as a result of the optimization. Because the objective function includes coefficients of loss, the obtained loss values will be smaller than the losses at the operating point (Sect. 3.5).

The other parts of the daily scheduling problem (start and stop constraints variable coupling constraints, supply constraints, fuel constraints) are determined on the basis of the preceding data taking into account the forecasted values of consumers' power demands on that particular day. Note that the forecasted demand deviates by at most 1–2 % from the actual demand.

5.2   Solution Approaches

The large-scale, mixed variable optimization model instance, generated as described previously and containing real and 0 or 1 -valued variables, can be solved numerically using different algorithms.

In the first line, application of the *Benders decomposition method* was considered; the whole model could have been solved using this algorithm in one major step. However, this method was soon rejected because it would have meant (Sect. 5.4) that a large number of integer programming problems had to be solved in each iteration of the decomposition (the solution of a single one of these subproblems would be a difficult task by itself). These subproblems, involving only integer variables, do not have the decomposition structure that is characteristic of the original large-scale model (after excluding some of the constraints).

The *branch-and-bound method* was also considered. In this case, the linear programming model arising in the algorithm could have been solved using the *Dantzig–Wolfe decomposition* algorithm. The reason this method was rejected was that the number of integer variables can be very high (up to 400), and this feature makes the branch-and-bound method practically inapplicable.

Relying on the foregoing considerations and on the physical background of the scheduling problem, a decomposition-based optimization method was chosen, including some heuristic elements as well. Roughly speaking, the problem is solved in a periodwise fashion, while the combination of an overall solution and the fulfillment of the fuel constraints are ensured by secondary assumptions. The procedure will be presented in the next section.

5.3   Optimization Method

The constraints corresponding to the simplified model of the scheduling problem are not taken into account in an explicit form (all constraints at one time) but separately, by constraint group. This decomposition is possible because the separate periods are connected by fuel constraints only, apart from the start-and-stop constraints.

The optimizing algorithm consists of the following steps (the algorithm is described for the case where there is only one fuel constraint):

1. The fuel constraint is omitted.
2. The resulting large-scale, mixed-variable programming problem (where the interconnections among the separate periods are ensured by the start-and-stop constraints and the mode-of-operation variables corresponding to the stagnation phases) is solved in the following way (the third and fourth steps of the algorithm).
3. The first, second, and third voltage check periods are solved in such a way that specific restrictions are imposed on the possible values of the mode-of-operation variables. In the solution of the first voltage check period, all the modes of operation applicable in that particular day may occur with no restrictions. In the solution of the second voltage check period (i.e., the first period of the first stagnation phase), only those modes of operation are permitted that can be achieved by shutdown actions from the modes of operation of the solution of the first voltage check period. Finally, those modes of operation can be applied in the third voltage check period (period of morning peak demand) that can be achieved by a start action from the modes of operation of the solution of the second voltage check period.
4. Afterward, the remaining problems corresponding to the remaining periods are solved in succession in such a way that the values of the mode-of-operation variables of the already solved periods preceding respectively following the given period are taken into account. This is done as follows.

   In a stop phase, only those modes of operation are considered in the solution of the given period that can be achieved

   (1) by a stop action from the system of modes of operation of the earlier, already solved, period of the stop phase (in short, by a stop action from the previous period) and from which

   (2) by a further stop action the system of modes of operation of the later, already solved, period of the stop phase can be obtained.

   Similarly, in a start phase only those modes of operation are allowed in the solution of a given period that can be achieved

   (1) by a start action from the system of modes of operation of the earlier, already solved, period of the start phase (in short, by a start action from a previous period) and

   (2) from which the system of modes of operation of a later, already solved, period in the start phase can be achieved by a further start action.

   Both at the stop and start actions, there can be (and are) nonchanging modes of operation variables.

   For example, in the solution of the second period only those modes of operation can be considered that can be achieved by a stop action from the first voltage check period and from which the modes of operation of the second (already solved) voltage check period can be achieved by stop actions only. In the solution of the third period the modes of operation of the second period and the second voltage check period should be considered, and so forth. For periods after the third voltage check period but preceding the second stagnation phase, only those modes of operation are permitted that depend on the periods solved earlier and can be achieved by stop actions, while at the solution of the periods following the stagnation phase, those modes of operation are allowed that can be achieved by start actions from the already solved periods.

The optimization problems corresponding to the specific periods are solved by the Benders decomposition method described in Sect. 5.4 that can take into account the previously described restrictions on the modes of operation.

5. Once the solutions to each of the periods of the day is determined, it is checked whether the overall solution satisfies the fuel constraint. If it does, the algorithm comes to an end; we have obtained an optimal solution. If it does not, an iterative procedure is applied to modify the solution achieved thus far. This procedure is described in the sixth step of the algorithm.

6. In the iterative procedure, a decrease respectively increase of the electric power production of the power plant with fuel constraint is essentially achieved by increasing respectively decreasing the fuel costs. A detailed description of this procedure follows.

According to (4.3.31) in Sect. 4.3.7, the fuel constraint can be formulated as

$$R_{i\,min} \leq h_i(\mathbf{p},\mathbf{x}) \leq R_{i\,max}, \tag{5.3.1}$$

where $(R_{i\,min} + R_{i\,max})/2$ is a predetermined, given value, and

$$\frac{R_{i\,max} + R_{i\,min}}{2} - R_{i\,min} = R_{i\,max} - \frac{R_{i\,max} + R_{i\,min}}{2} = \frac{k}{100}\frac{R_{i\,max} + R_{i\,min}}{2}$$

holds, where in this case k is a predetermined integer defining the permitted tolerance as a percentage (the usual value: $k = 3,4,5$). In the course of data preparation, it is necessary to ensure that with the given $R_{i\,min}$ and $R_{i\,max}$ constant values it is possible to find a pair of $\mathbf{p}$, $\mathbf{x}$ vectors satisfying inequality (5.3.1).

Let $(\mathbf{p_0},\mathbf{x_0})$ denote now the solution obtained in the third and fourth steps of the algorithm. If the inequality

$$R_{i\,min} \leq h_i(\mathbf{p_0},\mathbf{x_0}) \leq R_{i\,max}$$

does not hold (otherwise, optimization would have come to an end in the fifth step of the algorithm), the $\mathbf{x_0}$ vector will be fixed, i.e., no more changeovers take place in connection with the production modes. Let us consider the optimization problem that is obtained from the original problem by omitting the fuel constraint and substituting the vector $\mathbf{x_0}$. Let us denote this problem by $F_0(\mathbf{x_0})$; it is an ordinary linear programming problem consisting of 27 separate, mutually independent blocks, with $\mathbf{p}$ being an unknown vector. Starting with this problem, a sequence $F_1(\mathbf{x_0})$, $F_2(\mathbf{x_0})$... of problems will be generated, where the individual problems differ from one another only in the $c_1, c_2, \ldots, c_k$ coefficients in the objective function corresponding to the production-level variables of the active mode of operation at a power plant with a fuel constraint. The c_k coefficient appears in the problem $F_k(\mathbf{x}_0)$, while the coefficient corresponding to the original problem, denoted by c_0, appears in the problem $F_0(\mathbf{x}_0)$. Denoting the optimal solution of $F_k(\mathbf{x}_0)$ by $\mathbf{p}_k$, we determine the c_{k+1} coefficient, and with it the $F_{k+1}(\mathbf{x}_0)$ problem, by the following recursions ($s_0 = 1$):

$$c_{k+1} = c_k s_{k+1}/s_k \qquad (5.3.2)$$

$$s_{k+1} = \begin{cases} s_k \left[1 - 0.3 \dfrac{R_{i\,min} - h_i(\mathbf{p}_k, \mathbf{x}_0)}{\frac{R_{i\,min} + R_{i\,max}}{2}} \right] & \text{if } h_i(\mathbf{p}_k, \mathbf{x}_0) < R_{i\,min}, \\[3ex] s_k \left[1 - 0.3 \dfrac{R_{i\,max} - h_i(\mathbf{p}_k, \mathbf{x}_0)}{\frac{R_{i\,min} + R_{i\,max}}{2}} \right] & \text{if } h_i(\mathbf{p}_k, \mathbf{x}_0) > R_{i\,max}. \end{cases} \qquad (5.3.3)$$

To avoid unnecessary work, prices that have already been considered are excluded. This is achieved by prescribing bounds s_{min} and s_{max} for the s_k multiplier, where the bounds are themselves updated in the procedure, steadily narrowing the range of variation for s_k. Initially, let

$$s_{min} = 0.001 \qquad s_{max} = 1,000.$$

An s_k multiplier can cause underproduction respectively overproduction if $h_i(\mathbf{p}_k, \mathbf{x}_0) < R_{i\,min}$ [respectively $h_i(\mathbf{p}_k, \mathbf{x}_0) > R_{i\,max}$] holds in the case of the optimal model solution $\mathbf{p}_k$ formulated using the c_k coefficient that was calculated on the basis of s_k. If there is an underproduction in the case of the newly determined $\mathbf{p}_k$, then $s_{max} = s_k$ will be the new upper bound if $s_{min} < s_k < s_{max}$) holds (in the case of overproduction, $s_{min} = s_k$). Therefore, s_{max} comprises the smallest of the multipliers causing underproduction, while s_{min} contains the largest of the multipliers causing overproduction. If the inequality $s_{min} \leq s_k \leq s_{max}$ holds for the coefficient s_k calculated according to (5.3.3), the price is altered according to (5.3.2). Otherwise, the value of $s_k = \dfrac{s_{min} + s_{max}}{2}$ is used in the determination of c_k in (5.3.2). The s_{min} and s_{max} limits serve to constantly narrow the interval for choosing the possible multipliers; therefore, a c_k price that has already been used cannot reappear.

The sense of the described price modification is as follows. If at a power plant with a fuel constraint less electric power is generated than necessary [there is an underproduction, i.e., $h_i(\mathbf{p}_k, \mathbf{x}_0) < R_{i\,min}$ holds], then the coefficient of the objective function in the subsequent problem decreases compared to the previous one, so that the amount of electric power that is generated should increase (because the model aims at minimizing the costs). Similarly, in the case of overproduction [i.e., if $h_i(\mathbf{p}_k, \mathbf{x}_0) > R_{i\,max}$ holds], then the cost of electric power generation increases in the following $(k+1)^{\text{th}}$ problem, so that less electric power should be generated at the given power plant.

One of the following two cases holds with regard to the sequence $\mathbf{p}_0, \mathbf{p}_1, \ldots$ of optimal solutions of the optimization problems $F_0(\mathbf{x}_0), F_1(\mathbf{x}_0), F_2(\mathbf{x}_0) \ldots$:

a. A solution $\mathbf{p}_k$ is found for which (5.3.1) holds, i.e., the $(\mathbf{p}_k, \mathbf{x}_0)$ solution satisfies the fuel constraint as well;
b. A pair of solutions $\mathbf{p}_l$ and $\mathbf{p}_j$ is found where both $h_i(\mathbf{p}_l, \mathbf{x}_0) < R_{i\,\mathrm{min}}$ and $h_i(\mathbf{p}_j, \mathbf{x}_0) > R_{i\,\mathrm{max}}$ hold.

The occurrence of other cases (e.g., that of the $\mathbf{p}_k$ solutions causing underproduction only) is not possible, due to the verification of the data as described previously, following the discussion of formula (5.3.1). In case a, the optimization procedure terminates and $\mathbf{p}_k$ respectively $(\mathbf{p}_k, \mathbf{x}_0)$ is considered an optimal solution of the original problem. In case b, we proceed as follows. After fixing $\mathbf{x}_0$, the function $h_i(\mathbf{p}_l, \mathbf{x}_0)$ is linear in $\mathbf{p}$. Therefore, from the equality

$$\alpha h_i(\mathbf{p}_l, \mathbf{x}_0) + (1 - \alpha) h_i(\mathbf{p}_j, \mathbf{x}_0) = \frac{R_{i\,\mathrm{min}} + R_{i\,\mathrm{max}}}{2}$$

the multiplier $0 \le \alpha \le 1$ can be determined, and with it the vector $\mathbf{p}_{\mathrm{opt}} = \alpha \cdot \mathbf{p}_l + (1 - \alpha)\mathbf{p}_j$, which is considered an optimal solution since it is feasible, and

$$h_i(\mathbf{p}_{\mathrm{opt}}, \mathbf{x}_0) = \frac{R_{i\,\mathrm{min}} + R_{i\,\mathrm{max}}}{2}$$

holds as well.

Naturally, instead of α, the numbers α_1 and α_2 derived from the equalities

$$\alpha_1 h_i(\mathbf{p}_l, \mathbf{x}_0) + (1 - \alpha_1) h_i(\mathbf{p}_j, \mathbf{x}_0) = R_{i\,\mathrm{min}},$$

$$\alpha_2 h_i(\mathbf{p}_l, \mathbf{x}_0) + (1 - \alpha_2) h_i(\mathbf{p}_j, \mathbf{x}_0) = R_{i\,\mathrm{max}} \qquad (5.3.4)$$

may also be used, where α_1 and α_2 are the two values corresponding to the lower and upper bounds in the fuel constraint.

The application of α is motivated by the fact that the fuel constraint can also take the form of an equality by prescribing the consumption of a given quantity of fuel respectively the requirement to generate a given $(R_{i\,\mathrm{min}} + R_{i\,\mathrm{max}})/2$ amount of energy.

5.4 Benders Decomposition Method for Solving Subproblems

In the case of the optimization method presented in the preceding Sect. 5.3, the whole problem decomposes into 27 subproblems. On the basis of their types they can be arranged in three groups:

a. Problems corresponding to the voltage check periods;
b. Problems corresponding to the nonfirst periods of the stagnation phase;
c. Problems corresponding to the remaining so-called normal periods.

The problems in groups a and c are mixed-integer linear optimization problems, and the *Benders decomposition method*, utilizing the characteristics of these problems, is used to solve them. The problems in group b are linear programming problems with a single constraint and individual bounds on the variables; a greedy algorithm is used to solve them (Nemhauser and Wolsey, [52]).

In this section the *Benders decomposition method* and the simplifications carried out in its application are discussed, and only the mixed-integer linear case is considered. Readers interested in the details are encouraged to consult Benders [1], Lasdon [44], and Nemhauser and Wolsey, [52].

The *Benders decomposition* is used to solve optimization problems of the following type (the notations in the description of the algorithm and the problem are independent and different from the notations introduced in previous sections):

$$\mathbf{A}\mathbf{x} + \mathbf{F}\mathbf{y} \le \mathbf{b},$$

$$\mathbf{x} \ge \mathbf{0},$$

$$\mathbf{y} \in Y, \tag{5.4.5}$$

$$\max(\mathbf{c}^T\mathbf{x} + \mathbf{f}^T\mathbf{y}),$$

where: $\mathbf{A}$: $m \times n_1$ matrix;
$\mathbf{F}$: $m \times n_2$ matrix;
$\mathbf{c}, \mathbf{x}$: n_1-dimensional vector;
$\mathbf{f}, \mathbf{y}$: n_2-dimensional vector;
$\mathbf{b}$: m-dimensional vector;
Y: n_2-dimensional discrete set.

The problem to be solved is equivalent to the optimization problem

$$x_0 + ((\mathbf{u}^j)^T\mathbf{F} - \mathbf{f}^T)\mathbf{y} \le (\mathbf{u}^j)^T\mathbf{b}, \qquad j = 1,\dots,p,$$

$$(\mathbf{v}^j)^T\mathbf{F}\mathbf{y} \le (\mathbf{v}^j)^T\mathbf{b}, \qquad j = 1,\dots,r,$$

$$\mathbf{y} \in Y, \tag{5.4.6}$$

$$\max x_0,$$

where $\mathbf{u}^j$, $j = 1,\dots,p$, and $\mathbf{v}^j$, $j = 1,\dots,r$, are the extremal points respectively extremal directions of a polyhedron defined by

$$\mathbf{A}^T\mathbf{u} \ge \mathbf{c},$$

$$\mathbf{u} \ge \mathbf{0}. \tag{5.4.7}$$

Instead of the mixed-variables problem (5.4.5), the equivalent problem (5.4.6) will be solved; apart from the variable x_0, it is an integer programming problem. Relaxation is used to sole this latter problem because there is a large number of constraints that are not available in explicit form. The result is an iterative process

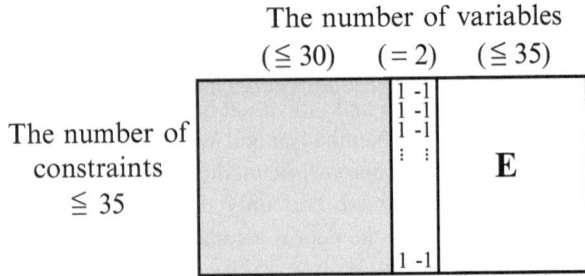

The number of variables

(≤ 30)   ($= 2$)   (≤ 35)

The number of
constraints
≤ 35

Fig. 5.1 Structure of the matrix of the first linear programming block to be solved in voltage check periods

where the ith iteration consists of the solution of both a relaxation (obtained by disregarding constraints) of problem (5.4.6) (let the optimal solution be denoted by $\mathbf{y}^i$) and the linear programming problem

$$\mathbf{A}^T\mathbf{u} \geq \mathbf{c},$$

$$\mathbf{u} \geq \mathbf{0},$$

$$\min(\mathbf{b} - \mathbf{F}\mathbf{y}^i)^T\mathbf{u}. \tag{5.4.8}$$

The feasible domains of problem (5.4.8) are the same during the iterations, only the objective function vector changes as a function of the optimal solution $\mathbf{y}^i$ of the previously solved relaxation of problem (5.4.6). The optimization problem (5.4.8) serve to check the optimality criterion and repeatedly generate constraints of the optimization problem (5.4.6).

At the end of the iterative procedure, the continuous ($\mathbf{x}$) part of the optimal solution of (5.4.5) can be determined without solving a further linear programming problem. Relying on duality theory (Prékopa [59]), the continuous part of the solution can be computed on the basis of the simplex tableau corresponding to the optimal solution of the last solved problem (5.4.8) (Hoffer [35]).

A further simplification is as follows: in the first iteration, the two-phase simplex method is used to solve problem (5.4.8) (Prékopa [59]), while in subsequent iterations the solution algorithm starts in the second phase, with the optimal solution from the previous iteration serving as the starting feasible solution.

The (5.4.8) linear programming problem, which occurs in the course of the solution of the problems corresponding to the voltage check periods, consists of two independent blocks. Therefore, its solution is decomposed into the solution of two separate problems (see Figs. 5.1 and 5.2 for the matrices of these two linear programming problems).

The application of another idea for the solution of the linear programming problem (5.4.8) corresponding to normal periods is in preparation. The application of the algorithm presented in [36] by Hoffer could significantly decrease the computational time needed to generate the optimal daily schedule.

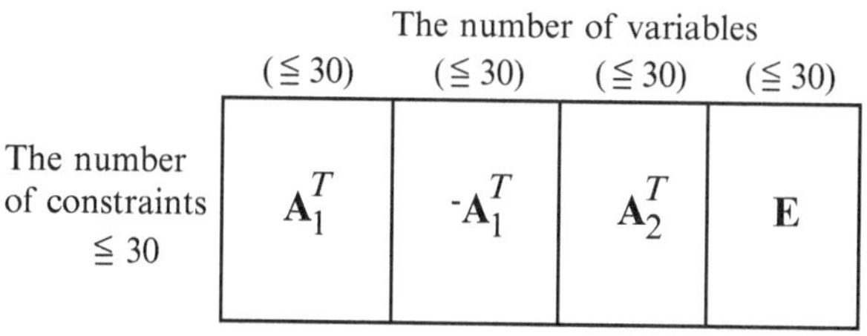

The number of variables

(≤ 30)	(≤ 30)	(≤ 30)	(≤ 30)
$\mathbf{A}_1^T$	$\mathbf{-A}_1^T$	$\mathbf{A}_2^T$	$\mathbf{E}$

The number of constraints ≤ 30

Fig. 5.2 Structure of the matrix of the second linear programming block to be solved in voltage check periods

The role of matrix **F** in the decomposition method is quite peculiar. It is only used in matrix multiplications. Multiplication from the right occurs in the computation of the constraints for the relaxations of (5.4.6), while multiplication from the left is used to calculate the coefficients of the objective function of problem (5.4.8). As the matrices corresponding to the mode-of-operation variables in the problems of both the normal and the voltage check periods are specially structured, apart from the submatrix of branch-load constraints, the multiplication can be easily expressed explicitly in terms of the nonzero elements of the matrices involved. Utilizing this fact, matrices in the program are not represented by filling them up with their elements, but the multiplicative operations are carried out in accordance with the characteristics of the special structure.

The fulfillment of conditions related to mode-of-operation variables (describing the characteristics of set Y) has been implemented within the enumeration type algorithm applied tp the solution of the relaxed problem (5.4.6) as follows:

- To take into account the systems of modes of operation of a previous and a subsequent period: fixing the relevant mode of operation variables on the levels of 0 or 1;
- To comply with further prescriptions for a given period: the compulsory operation or standstill of certain modes of operation by adequately fixing the mode-of-operation variables;
- To satisfy conditions (4.3.19) describing the special logics of mode-of-operation variables: by analyzing all the consequences of fixing and tying in an implicit enumeration algorithm and fixing the relevant variables, e.g., if during enumeration a mode-of-operation variable is assigned a value of 1, then mode-of-operation variables of a smaller index belonging to the same power plant are fixed at a value of 1. If the same variable is assigned a value of 0, then the relevant mode-of-operation variables of a higher index would be fixed at a value of 0.

Appendix
The Transmission Network of Electric Power Systems

A.1 Mathematical Model of Electric Networks

In this section we build a mathematical model of electric networks. Our starting point consists of the models presented in the paper [41] and in the books [63] and [70]. We modify these models since the construction of a mathematical model of a transmission network requires a new systematic analysis of the known facts and consideration of certain facts from a new angle. For instance, we introduce a new admittance transformation; see [50]. In our opinion, this facilitates a method of analysis that is far less complicated than that based on the transformation in the literature and it fits better into the conceptual framework of physics.

The electric network is modeled by a directed graph, and for this reason we summarize some basic facts from graph theory.

Let us consider a finite set $\mathcal{N}$ and another set $\mathcal{A}$, the latter consisting of pairs of elements from $\mathcal{N}$. The sets $\mathcal{N}$ and $\mathcal{A}$ together are called a graph, while the elements of set $\mathcal{N}$ are called *nodes* and the elements of set $\mathcal{A}$ are called *edges*. The pairs of elements included in set $\mathcal{A}$ can either be ordered (meaning that we specify which of the elements is considered as first and which as second) or unordered. In the former case the edges are called *directed* and in the latter case they are called *undirected*.

A graph with undirected edges is called an undirected graph, while a graph with directed edges is a directed graph. To distinguish them in the notation, the set of edges of the latter will be denoted by $\mathcal{A}^D$. For the two types of graphs we introduce the symbols $(\mathcal{N}, \mathcal{A})$ for the undirected case and $(\mathcal{N}, \mathcal{A}^D)$ for the directed case.

First undirected graphs are discussed. If set $\mathcal{N}$ has n elements, then the numbers $1, \ldots, n$ are used to denote the nodes of the graph, and every edge is a pair of distinct numbers $\{i, j\}$, i.e., a subset consisting of two elements of the set $\mathcal{N} = \{1, \ldots, n\}$. Edge $\{i, j\}$ is interpreted as connecting nodes i and j.

Obviously, if the number of nodes is n, the number of edges is at most $n(n-1)/2$.

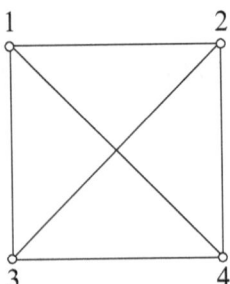

Fig. A.1 Fourth example of undirected graph

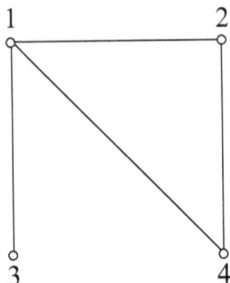

Fig. A.2 First example of undirected graph

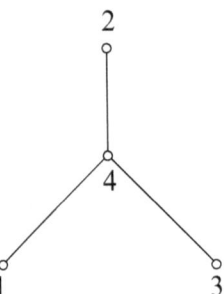

Fig. A.3 Second example of undirected graph

A graph can be visualized by choosing n different points on the plane representing the nodes of the graph, and points i, j are connected by a line if $\{i, j\} \in \mathscr{A}$. The following figures depict four graphs.

Some important notions are introduced next, followed by some theorems along with their proofs.

Subgraph.   Graph $\mathscr{G}' = (\mathscr{N}', \mathscr{A}')$ is called a subgraph of graph $\mathscr{G} = (\mathscr{N}, \mathscr{A})$ if both $\mathscr{N}' \subseteq \mathscr{N}$ and $\mathscr{A}' \subseteq \mathscr{A}$ hold.

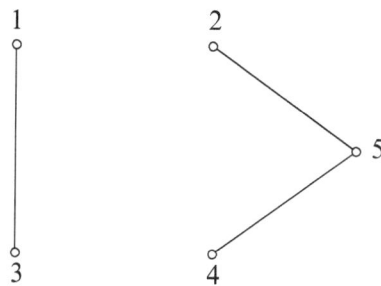

Fig. A.4 Third example of undirected graph

Path. If $s, i_1, \ldots, i_k, t$ are different elements of the set $\mathcal{N}$, then the sequence of edges $\{s, i_1\}, \{i_1, i_2\}, \ldots, \{i_{k-1}, i_k\}, \{i_k, t\}$ is called a path connecting nodes s and t.

Circuit. If $s, i_1, \ldots, i_k$ are different elements of set $\mathcal{N}$, then the sequence of edges $\{s, i_1\}, \{i_1, i_2\}, \ldots, \{i_{k-1}, i_k\}, \{i_k, s\}$ is called a circuit.

Connected graph. If any two nodes of a graph can be connected by a path, then the graph is called a connected graph. (The graph in Fig. A.1 is not connected, whereas those in Figs. A.2–A.4 are connected.)

Tree. A graph is called a tree if it is connected and contains no circuit. (The graph in Fig. A.4 is a tree.)

Spanning tree. Subgraph $(\mathcal{N}', \mathcal{A}')$ of graph $(\mathcal{N}, \mathcal{A})$ is called a spanning tree if $(\mathcal{N}', \mathcal{A}')$ is a tree and $\mathcal{N} = \mathcal{N}'$ holds.

Isolated node. If node i does not belong to any of the edges, then it is called an isolated node.

Terminal node. If node i belongs to exactly one edge, then it is called a terminal node of the graph.

Theorem 1 *A tree with n nodes has n − 1 edges.*

Proof. The assertion is obviously true in the case of $n = 1$. (A graph having a single node is connected and circuit free, and thus it is a tree).

In the case of $n \geq 2$, it is first proved that the tree has a terminal node. In fact, the two nodes that are connected by the longest possible path are terminal since otherwise the path could be extended.

In the case of $n \geq 2$, the assertion can be proved by mathematical induction. For $n = 2$ it is obviously true. Presupposing that the proposition is true in the case of any tree with n nodes, let us consider a tree of $n + 1$ nodes, where $n \geq 2$. Dropping an arbitrary terminal node and the adjacent edge of the tree, we get a tree of n nodes that has $n - 1$ edges, according to the induction hypothesis. Consequently, together with the edge that has just been dropped, in the original tree we have n edges, and the assertion in the case of $n + 1$ nodes is proved. Thus, Theorem 1 is proved. □

Theorem 2 *Every pair of tree nodes can be connected by one and only one path.*

Proof. The definition of a tree requires the existence of a path. If there were a pair of nodes that could be connected by two different paths, then the graph would contain a circuit as well. This is not possible, however, because a tree is circuit free. The proof is complete.   □

Theorem 3 *If every pair of nodes of a graph can be connected by one and only one path, then the graph is a tree.*

Proof. Because the graph is connected according to the assumption, only its circuit-free character must be proved. This can easily be seen since any two nodes of a circuit can be connected by two different paths using the edges in the circuit. Theorem 3 is proved.   □

The preceding theorems can be summarized in one theorem, as follows.

Theorem 4 *For a graph $\mathcal{G}$ with n nodes, the following statements are equivalent:*

(a) *Graph $\mathcal{G}$ is a tree;*
(b) *Every pair of nodes of graph $\mathcal{G}$ can be connected by a unique path;*
(c) *Graph $\mathcal{G}$ is connected and has $n - 1$ edges;*
(d) *Graph $\mathcal{G}$ is circuit free and has $n - 1$ edges.*

Theorem 5 *If the graph $\mathcal{G} = (\mathcal{N}, \mathcal{A})$ is a tree and $\{i, j\} \notin \mathcal{A}$, then there exists precisely one circuit in the graph $\mathcal{G}_1 = (\mathcal{N}, \mathcal{A} \cup \{i, j\})$.*

Proof. There must be a circuit in graph $\mathcal{G}_1$ because it has as many nodes as $\mathcal{G}$, though there is an additional edge in it. Consequently, graph $\mathcal{G}_1$ is not a tree but it is connected; consequently, it contains a circuit. Now it remains to prove that there do not exist two different circuits in graph $\mathcal{G}_1$. Using an indirect proof, let us suppose that there are two different circuits in $\mathcal{G}_1$. In this case, edge $\{i, j\}$ must belong to both of them since $\mathcal{G}$ contains no circuit. If edge $\{i, j\}$ is deleted in both circuits, then there are two different paths connecting nodes i and j. This is not possible according to Theorem 2.   □

Theorem 6 *If graph $\mathcal{G} = (\mathcal{N}, \mathcal{A})$ is connected, then it contains a spanning tree.*

Proof. A constructive proof follows. Let $\mathcal{N}_1 = \{i_1\}$, $\mathcal{A}_1 = \emptyset$, where $i_1 \in \mathcal{N}$ is arbitrary. Then $\mathcal{G}_1 = (\mathcal{N}_1, \mathcal{A}_1)$ is a subgraph of $\mathcal{G}$, and it is trivially a tree. Let us assume that the subgraph $\mathcal{G}_k = (\mathcal{N}_k, \mathcal{A}_k)$ of graph $\mathcal{G}$ is given and that it is a tree, $1 \le k \le n - 1$. Then $\mathcal{G}_{k+1}$ can be constructed in the following way. Because $\mathcal{G}$ is connected, there exist $i_p \in \mathcal{N}_k$ and $i_q \in \mathcal{N} \setminus \mathcal{N}_k$ for which $\{i_p, i_q\} \in \mathcal{A}$ holds. Let $\mathcal{N}_{k+1} = \mathcal{N}_k \cup \{i_q\}$ and $\mathcal{A}_{k+1} = \mathcal{A}_k \cup \{i_p, i_q\}$; then it is clear that $\mathcal{G}_{k+1} = (\mathcal{N}_{k+1}, \mathcal{A}_{k+1})$ will also be a tree, and therefore $\mathcal{G}_n = (\mathcal{N}_n, \mathcal{A}_n)$ is a spanning tree. The proof is complete.   □

The edges of an undirected graph have been identified up till now with a pair of nodes formed by the endpoints of the relevant edge. For the modeling of the electric network, however, a notion of an undirected graph is needed where two nodes may be connected by more than one edge. Therefore, the edges can no longer

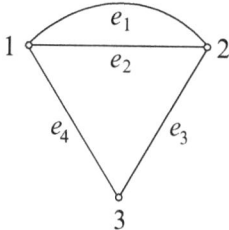

Fig. A.5 Undirected graph with parallel edges

be identified with the pair of nodes of endpoints. For this reason, the definition of the undirected graph needs to be generalized. In the sequel, an undirected graph will be defined as follows.

Let us consider the finite sets $\mathcal{N}$, $\mathcal{A}$, and let us assign to each element of $\mathcal{A}$ an unordered pair of elements of the set $\mathcal{N}$ (more precisely, a two-element subset of $\mathcal{N}$). The sets $\mathcal{N}$, $\mathcal{A}$, together with the preceding assignments, are called an undirected graph and are still denoted by the symbol $(\mathcal{N}, \mathcal{A})$ for the sake of simplicity. The elements of set $\mathcal{N}$ are called nodes, while the elements of set $\mathcal{A}$ are called edges.

Let the number of elements of $\mathcal{N}$ be n and the number of elements of $\mathcal{A}$ be m. The nodes of the graph can be denoted also in this case by the numbers $1,\ldots,n$. The edges of the graph are denoted by $e_1, e_2, \ldots, e_m$. If $e_k \in A$ and the pair of elements attached to the edge is $\{i, j\}$, this is denoted by the symbol $e_k \to \{i, j\}$. If $e_k \to \{i, j\}$ and $e_l \to \{i, j\}$, $k \neq l$, hold, then edges e_k and e_l are called parallel edges.

Obviously, the undirected graph as defined at the beginning of this section is a special case of this new definition: it corresponds to the prescription requiring that different pairs of nodes should be assigned to different edges.

The visualization of the graph can be done in a similar way to what was done previously. The n points in a plane represent the nodes of the graph, and points $\{i, j\}$ are connected by as many lines as there are edges assigned to the pair of numbers $\{i, j\}$. The edges are visualized along with their identifiers. In the graph in Fig. A.5, e_1 and e_2 are parallel edges.

The definitions of a subgraph, path, and circuit are modified in the following way.

Subgraph.   A graph $\mathcal{G}' = (\mathcal{N}', \mathcal{A}')$ is called a subgraph of graph $\mathcal{G} = (\mathcal{N}, \mathcal{A})$ if $\mathcal{N}' \subseteq \mathcal{N}$, if $\mathcal{A}' \subseteq \mathcal{A}$, and, furthermore, if the same pairs of nodes are assigned to the elements of set $\mathcal{A}'$ in $\mathcal{G}'$ as in graph $\mathcal{G}$.

Path.   If $s, i_1, \ldots, i_k$ are different elements of set $\mathcal{N}$, then the sequence of edges $e_{j_1}, \ldots, e_{j_{k+1}}$ is called a path connecting nodes s and t if $e_{j_1} \to \{s, i_1\}$, $e_{j_2} \to \{i_1, i_2\}, \ldots, e_{j_k} \to \{i_{k-1}, i_k\}$, $e_{j_{k+1}} \to \{i_k, t\}$ are satisfied.

Circuit.   If $s, i_1, \ldots, i_k$ are different elements of set $\mathcal{N}$, the sequence of edges $e_{j_1}, \ldots, e_{j_{k+1}}$ is called a circuit if $e_{j_1} \to \{s, i_1\}$, $e_{j_2} \to \{i_1, i_2\}, \ldots, e_{j_k} \to \{i_{k-1}, i_k\}$, $e_{j_{k+1}} \to \{i_k, s\}$ hold.

All of the other definitions of undirected graphs as presented at the beginning of this section can be transferred verbatim to the generalized notion of undirected graphs by substituting the altered notions of a subgraph, path, and circuit as defined previously. The reader can easily check that all the theorems will be valid in the case of the more general graphs as well. The same proofs can be applied with minor modifications in some places.

In the sequel we will discuss directed graphs. If the directed graph $(\mathcal{N}, \mathcal{A}^D)$ has n nodes, they can be denoted by $1, \ldots, n$ in this case as well. For directed graphs every edge corresponds to an ordered pair (i, j) of numbers. We assume that $\mathcal{A}^D \subseteq \mathcal{N} \times \mathcal{N}$ holds, which means that any ordered pair of numbers corresponds to at most one edge. Consequently, for the denotation of the edges we may employ the corresponding ordered pairs of numbers. Let the number of edges be m, and let serial numbers be assigned to the individual edges. We assume that $(\mathcal{N}, \mathcal{A}^D)$ contains no loops, i.e., $(i, i) \notin \mathcal{A}^D$, $i = 1, \ldots, n$ holds.

Let $(\mathcal{N}, \mathcal{A})$ denote an undirected graph corresponding to $(\mathcal{N}, \mathcal{A}^D)$, which can be obtained from $(\mathcal{N}, \mathcal{A}^D)$ by considering the edges of this graph as being undirected.

To apply a simplified notation, the following notation is allowed. If the components of $\mathbf{d} \in \mathbb{R}^m$ (or $\mathbf{d} \in \mathbb{C}^m$) represent edge characteristics and (i, j) is assigned to edge l, then the notation $d_{i,j}$ will be permitted along with d_l to denote the lth component of vector $\mathbf{d}$. Furthermore, let us consider those edges of $(\mathcal{N}, \mathcal{A})$ one of whose endpoints is node i, $i = 1, \ldots, n$. We will denote by $J(i)$ the set of serial numbers of the other endpoints. The set of these edges in graph $(\mathcal{N}, \mathcal{A}^D)$ will be called the set of edges connected to node i. Finally, $i(l)$ denotes the serial number of the other endpoint of the edge with serial number l connected to the ith node.

Definition The directed graph $(\mathcal{M}, \mathcal{B}^D)$ is called a subgraph of the directed graph $(\mathcal{N}, \mathcal{A}^D)$ if $\mathcal{M} \subseteq \mathcal{N}$ and $\mathcal{B}^D \subseteq \mathcal{A}^D$ hold.

Definition The directed graph $(\mathcal{N}, \mathcal{A}^D)$ is connected if $(\mathcal{N}, \mathcal{A})$ is connected.

In the sequel we will consider connected graphs only, i.e., it is assumed that $(\mathcal{N}, \mathcal{A}^D)$ is a connected directed graph.

Definition Let $(\mathcal{N}, \mathcal{F}^D)$ be a subgraph of graph $(\mathcal{N}, \mathcal{A}^D)$. $(\mathcal{N}, \mathcal{F}^D)$ is called a spanning tree if in the undirected graph $(\mathcal{N}, \mathcal{A})$, $(\mathcal{N}, \mathcal{F})$ is a spanning tree.

The elements of $\mathcal{A}^D \setminus \mathcal{F}^D$ are called linking edges, and their number is denoted by k. As a result of Theorem 1, the number of edges of $(\mathcal{N}, \mathcal{F}^D)$ is $n - 1$, and $k = m - n + 1$ holds.

Definition Subgraph $(\mathcal{N}_H, \mathcal{H}^D)$ of graph $(\mathcal{N}, \mathcal{A}^D)$ is called a circuit if $(\mathcal{N}_H, \mathcal{H})$ is a circuit in graph $(\mathcal{N}, \mathcal{A})$.

Circuits may be endowed with an orientation, that is, the fixing of a so-called traversal of $(\mathcal{N}_H, \mathcal{H})$, intuitively. If there are more than two nodes in a circuit, then, with respect to these nodes, a cyclic ordering is fixed. Then, in the case of $(i, l) \in \mathcal{H}^D$, edge (i, l) will be called positively oriented, with respect to the circuit, if in the cyclic order the order of the nodes is i, l. Otherwise, (i, l) will be called negatively oriented. If there are only two nodes in the circuit, then the circuit takes the form $\mathscr{H}^D = \{(i, l), (l, i)\}$, and the exact definition of *traversal* can easily be formulated in an analogous way to the preceding case.

Definition Let $\mathscr{N}_1, \mathscr{N}_2$ be a partition of set $\mathscr{N}$, i.e., $\mathscr{N}_1 \cup \mathscr{N}_2 = \mathscr{N}$ and $\mathscr{N}_1 \cap \mathscr{N}_2 = \emptyset$ should hold. The following subset of $\mathscr{A}^D$ will be called a cut set corresponding to this partition and will be denoted in this section by $C(\mathscr{N}_1, \mathscr{N}_2)$:

$$C(\mathscr{N}_1, \mathscr{N}_2) = \{(i, l) \mid (i, l) \in \mathscr{A}^D, \quad i \in \mathscr{N}_1, l \in \mathscr{N}_2, \quad \text{or} \quad i \in \mathscr{N}_2, l \in \mathscr{N}_1\}.$$

The cut sets can have orientations, meaning that one possible order of $\mathscr{N}_1, \mathscr{N}_2$ is fixed, that is, for example, the $(\mathscr{N}_1, \mathscr{N}_2)$ pair of sets is handled as an ordered pair. In the case of $(i, l) \in C(\mathscr{N}_1, \mathscr{N}_2)$, this edge is considered positively oriented with respect to the cut set if $i \in \mathscr{N}_1$ and negatively oriented if $i \in \mathscr{N}_2$ holds.

Definition The cut set $C(\{l\}, \mathscr{N} \setminus \{l\})$ is called a node cut set belonging to node l for $l = 1, \ldots, n$. Obviously, $C(\{l\}, \mathscr{N} \setminus \{l\})$ is the set of edges connected to node l.

For the sake of simplicity of presentation, a spanning tree $(\mathscr{N}, \mathscr{F}^D)$ is chosen from graph $(\mathscr{N}, \mathscr{A}^D)$ and is regarded as fixed in the sequel.

Definition The basic circuits corresponding to the links $\mathscr{A}^D \setminus \mathscr{F}^D$ are the following circuits: let $(i, l) \in \mathscr{A}^D \setminus \mathscr{F}^D$. As a result of Theorem 5, graph $(\mathscr{N}, \mathscr{F}^D \cup \{(i, l)\})$ contains one and only one circuit, which is called the basic circuit generated by the edge (i, l).

Therefore, the number of basic circuits corresponding to the tree $(\mathscr{N}, \mathscr{F}^D)$ is k.

Definition By the basic cut sets belonging to the edges of tree $(\mathscr{N}, \mathscr{F}^D)$, the following cut sets are meant: let $(i, l) \in \mathscr{F}^D$. If edge (i, l) is deleted in the tree, the tree disintegrates into two disjoint trees. Regarding the nodes of these trees, a partition of $\mathscr{N}$ results. The cut set in $(\mathscr{N}, \mathscr{A}^D)$, corresponding to this partition, is called a basic cut set generated by the edge (i, l).

On the basis of the definition and of Theorem 1, it can be seen that the number of basic cut sets corresponding to the tree $(\mathscr{N}, \mathscr{F}^D)$ is $n - 1$.

In the sequel we will need various incidence matrices describing the structure of graph $(\mathscr{N}, \mathscr{A}^D)$ and the orientations of cut sets and circuits. It is assumed that cut sets and circuits corresponding to a graph are endowed with serial numbers and that the basic cut sets, basic circuits, and node cut sets are also provided with serial numbers corresponding to the resulting ordering. The definitions of the various incidence matrices are as follows.

$\hat{Q}$:   A cut-set incidence matrix describing cut sets and their fixed orientation. Its size is $p \times m$, where p is the number of cut sets corresponding to graph $(\mathscr{N}, \mathscr{A}^D)$.

Definition:

$$\hat{Q}_{il} = \begin{cases} 1 \text{ if edge } l \text{ belongs to cut set } i \text{ and it is} \\ \quad \text{positively oriented with respect to the cut set;} \\ -1 \text{ if edge } l \text{ belongs to cut set } i \text{ and it is} \\ \quad \text{negatively oriented with respect to the cut set;} \\ 0 \text{ otherwise.} \end{cases} \quad \text{(A.1.1)}$$

Q: Reduced cut-set incidence matrix of size $(n-1) \times m$. For its definition we assume that the cut sets are provided with the following orientations: the orientations of basic cut sets are of the sort that the edges generating the cut set are positively oriented, while the orientations of the remaining cut sets are arbitrarily fixed. Then **Q** is the submatrix formed by choosing those $n-1$ rows of $\hat{\mathbf{Q}}$ that correspond to the basic cut sets.

Â: Node-edge incidence matrix of size $n \times m$. It is assumed that the cut sets of $(\mathcal{N}, \mathcal{A}^D)$ are endowed with the following orientations: the orientations of the node cut sets are of the sort that with respect to them, the edges emanating from the node are positively oriented; the orientation of the remaining cut sets is arbitrarily fixed. Then $\hat{\mathbf{A}}$ is a submatrix of $\hat{\mathbf{Q}}$ corresponding to the rows of the n node cut sets. Because the node-edge incidence matrix will be needed frequently later on, its explicit form is also provided:

$$\hat{A}_{il} = \begin{cases} 1 \text{ if the } l\text{th edge is connected to the } i\text{th node and the node} \\ \quad \text{is the starting point of the edge;} \\ -1 \text{ if the } l\text{th edge is connected to the } i\text{th node and the node} \\ \quad \text{is the endpoint of the edge;} \\ 0 \text{ otherwise.} \end{cases} \quad \text{(A.1.2)}$$

A: Reduced node-edge incidence matrix of size $(n-1) \times m$. This matrix is formed by deleting one of the rows of $\hat{\mathbf{A}}$. In this way, n reduced node-edge incidence matrices can be associated with $\hat{\mathbf{A}}$.

B̂:   Circuit-edge incidence matrix, describing both the circuits corresponding to graph $(\mathcal{N}, \mathcal{A}^D)$, and their fixed orientations. Its size is $h \times m$, where h is the number of circuits in the graph.

Definition:

$$\hat{B}_{il} = \begin{cases} 1 \text{ if the } l\text{th edge is in the } i\text{th circuit and with respect to the} \\ \quad \text{circuit it is positively oriented;} \\ -1 \text{ if the } l\text{th edge is in the } i\text{th circuit and with respect to the} \\ \quad \text{circuit it is negatively oriented;} \\ 0 \text{ otherwise.} \end{cases} \quad \text{(A.1.3)}$$

B: Reduced circuit-edge incidence matrix of size $k \times m$. For this definition the circuits are oriented in the following way. The orientation of the basic circuits are of the sort that the edge generating the circuit is of positive orientation, while the orientation of the remaining circuits is arbitrarily fixed. Then $\mathbf{B}$ is the submatrix of matrix $\hat{\mathbf{B}}$, corresponding to the rows of the basic circuits.

For the sake of simplicity of notation, let us assume that the serial numbers of the edges of the graph start with the link edges corresponding to the fixed tree $(\mathcal{N}, \mathcal{F}^D)$. Then $\mathbf{B}$ and $\mathbf{Q}$ take the following so-called normal form:

$$\mathbf{B} = (\mathbf{E}_k, \mathbf{F}); \qquad \mathbf{Q} = (\mathbf{Q}', \mathbf{E}_{n-1}), \tag{A.1.4}$$

where $\mathbf{E}_k$, $\mathbf{E}_{n-1}$ are unit matrices of appropriate sizes.

Later on the following theorem will be needed; its proof can be found in [63], for example.

Theorem 7 *The directed graph* $(\mathcal{M}, \mathcal{B}^D)$ *contains a circuit if and only if the columns of its node-edge incidence matrix* $\hat{\mathbf{A}}$ *are linearly dependent.*

The next theorem concerns the rank of various incidence matrices. For its formulation let us introduce the notation $r(\mathbf{D})$ for the rank of a matrix, where $\mathbf{D}$ is an arbitrary matrix.

Theorem 8 *For the incidence matrices of the connected graph* $(\mathcal{N}, \mathcal{A}^D)$ *the following assertions hold:*

(i) $r(\mathbf{Q}) = n - 1$, $r(\mathbf{B}) = k$;
(ii) $r(\hat{\mathbf{A}}) = n - 1$;
(iii) *If any row of* $\hat{\mathbf{A}}$ *is deleted, then the rows of the resulting reduced node-edge incidence matrices are linearly independent.*

Proof. Statement (i), on the basis of (A.1.4), is trivial. To prove assertion (ii), let us note first that if we add up the row vectors of $\hat{\mathbf{A}}$, then a zero vector results. Consequently, the rows of $\hat{\mathbf{A}}$ are linearly dependent, implying that $r(\hat{\mathbf{A}}) \leq n - 1$ holds. At the same time, it is possible to choose $n - 1$ linearly independent column vectors of $\hat{\mathbf{A}}$. For this it is sufficient to choose the column vectors corresponding to a spanning tree whose existence follows from Theorem 6. These vectors are in fact linearly independent on the basis of Theorem 7. Finally, the proof of (iii) easily follows from the fact that, due to a statement at the beginning of the proof of (ii), any row of $\hat{\mathbf{A}}$ can be expressed as a linear combination of the others. Thus, the proof is complete.   □

The next theorem describes an orthogonality relation.

Theorem 9 *For arbitrary orientations of the cut sets and circuits of graph* $(\mathcal{N}, \mathcal{A}^D)$, *the following relation holds:*

$$\hat{\mathbf{Q}}\hat{\mathbf{B}}^T = \mathbf{0}. \tag{A.1.5}$$

Proof. Let us choose the sth row in matrix $\hat{\mathbf{Q}}$ and the tth row in matrix $\hat{\mathbf{B}}$. This means that the sth cut set and the tth circuit are considered. If the cut set and the circuit have no common edges, then the scalar product is trivially zero. Otherwise, it is easy to see that the cut set and the circuit have an even number of common edges. Taking a traversal corresponding to the orientation of the circuit, the orientations of common edges with respect to the circuit and to the cut set are identical or opposite, in an alternating fashion. From this fact the proposition easily follows.   □

In the sequel let $\mathbb{C}^m$ denote the linear space of complex m-tuples over $\mathbb{C}$, endowed with the usual scalar product: for $\mathbf{a}, \mathbf{b} \in \mathbb{C}^m$, $\langle \mathbf{a}, \mathbf{b} \rangle = \sum_{i=1}^{m} a_i b_i^*$, where $*$ denotes the complex conjugate.

If $\mathbb{D}$ is a subspace of $\mathbb{C}^m$, its dimension will be denoted by $\dim \mathbb{D}$.

Because the elements of the matrices in Theorems 8 and 9 are real numbers, the assertions concerning ranks remain valid when taking $\mathbb{C}^m$ as the basis, and the orthogonality relation in (A.1.5) can also be formulated with scalar products in $\mathbb{C}^m$.

The following notations are introduced.

Let $\mathbb{C}_Q$ be the subspace of $\mathbb{C}^m$ spanned by the row vectors of $\hat{\mathbf{Q}}$, and let $\mathbb{C}_B$ be the subspace spanned by the row vectors of $\hat{\mathbf{B}}$. Since the rows of matrices $\hat{\mathbf{Q}}$ and $\hat{\mathbf{B}}$, corresponding to different orientations of cut sets and circuits, differ in the multiplication by (-1) at most, $\mathbb{C}_Q$ and $\mathbb{C}_B$ do not depend on the orientations of cut sets and circuits.

Theorem 10 *The following statements hold:*

(i) $\dim \mathbb{C}_Q = n - 1$;
(ii) $\dim \mathbb{C}_B = k$;
(iii) $\mathbb{C}_Q$ *and* $\mathbb{C}_B$ *are orthogonal complementary subspaces.*

Proof. Choosing appropriate orientations of the cut sets and circuits, the rows of matrix $\mathbf{Q}$ can be found among the rows of matrix $\hat{\mathbf{Q}}$, and the rows of matrix $\mathbf{B}$ can be found among the rows of matrix $\hat{\mathbf{B}}$. Therefore, on the basis of Theorem 8 we obtain

$$\dim \mathbb{C}_Q \geq n - 1, \qquad \dim \mathbb{C}_B \geq k. \qquad (A.1.6)$$

On the other hand, on the basis of (A.1.5), it can be seen that $\mathbb{C}_Q$ and $\mathbb{C}_B$ are orthogonal subspaces of $\mathbb{C}^m$. Therefore,

$$\dim \mathbb{C}_Q + \dim \mathbb{C}_B \leq m \qquad (A.1.7)$$

holds. With the help of the relation $m = n - 1 + k$, we have obtained the proofs of (i) and (ii). The orthogonality of the subspaces, along with (i) and (ii), imply (iii). Thus, the proof is complete.   □

The sequence of propositions related to the rank of the incidence matrices can be completed by applying our theorem.

Theorem 11 *For the connected graph* $(\mathcal{N}, \mathscr{A}^D)$, *with an arbitrary orientation of the cut sets and circuits, the following relations hold:* $r(\hat{\mathbf{Q}}) = n - 1$ *and* $r(\hat{\mathbf{B}}) = k$.

Proof. THe proof is trivial on the basis of Theorem 10.   □

Note that utilizing the orthogonality relation (A.1.5), $\mathbf{Q}' = -\mathbf{F}^T$ holds in relations (A.1.4). Therefore, $\mathbf{B}$ and $\mathbf{Q}$ take the following form:

$$\mathbf{B} = (\mathbf{E}_k, \mathbf{F}), \qquad \mathbf{Q} = (-\mathbf{F}^T, \mathbf{E}_{n-1}). \tag{A.1.8}$$

According to the fixed numbering of edges corresponding to relations (A.1.4), the partition of the components of vectors in $\mathbb{C}^m$ is as follows.

If $\mathbf{d} \in \mathbb{C}^m$, then $\mathbf{d} = \begin{pmatrix} \mathbf{d}_K \\ \mathbf{d}_F \end{pmatrix}$, where $\mathbf{d}_K \in \mathbb{C}^k$, $\mathbf{d}_F \in \mathbb{C}^{n-1}$.

In the sequel we discuss a description of the electrical states of networks.

Definition The following subspace of $\mathbb{C}^m$, denoted by $\mathscr{I}$, is called the current space:

$$\mathscr{I} = \{\mathbf{i} \mid \hat{\mathbf{Q}}\mathbf{i} = \mathbf{0}, \, \mathbf{i} \in \mathbb{C}^m\}. \tag{A.1.9}$$

The physical background is the following form of *Kirchhoff's first law*: in an electric network, the edge currents are of the sort that their signed sum is zero with regard to any oriented cut set of the network.

From Theorems 8 and 11 it follows that $\mathscr{I}$ is a k-dimensional subspace of $\mathbb{C}^m$, which can be specified in the following equivalent forms:

$$\mathscr{I} = \{\mathbf{i} \mid \mathbf{Q}\mathbf{i} = \mathbf{0}, \, \mathbf{i} \in \mathbb{C}^m\}, \tag{A.1.10}$$

$$\mathscr{I} = \{\mathbf{i} \mid \hat{\mathbf{A}}\mathbf{i} = \mathbf{0}, \, \mathbf{i} \in \mathbb{C}^m\}, \tag{A.1.11}$$

$$\mathscr{I} = \{\mathbf{i} \mid \mathbf{A}\mathbf{i} = \mathbf{0}, \, \mathbf{i} \in \mathbb{C}^m\}. \tag{A.1.12}$$

Of the preceding relations, (A.1.11) is the mathematical formulation of the following form of *Kirchhoff's first law*: the edge currents of electric networks are of the sort that the sum of the currents flowing inward equals the sum of the currents flowing outward with regard to any node of the network.

Equations (A.1.10) and (A.1.8) imply that the system of equations defining the current space can also be formulated in the following form:

$$-\mathbf{F}^T\mathbf{i}_K + \mathbf{i}_F = \mathbf{0}. \tag{A.1.13}$$

Further on, the components of the vectors $\mathbf{i} \in \mathscr{I}$ will also be called *edge currents*.

It can be seen from (A.1.13) that for arbitrarily chosen edge currents for the links the edge currents corresponding to the edges of the tree are uniquely determined.

Definition The following subspace of $\mathbb{C}^m$, denoted by $\mathscr{V}$, is called a voltage space:

$$\mathscr{V} = \{\mathbf{v} \mid \hat{\mathbf{B}}\mathbf{v} = \mathbf{0}, \, \mathbf{v} \in \mathbb{C}^m\}. \tag{A.1.14}$$

Considering the physical background, the definition corresponds to *Kirchhoff's second law*. This means that the edge voltages in electric networks are of the sort that their signed sum is zero for any oriented circuit of the network.

From Theorems 8 and 11 it follows that the dimension of $\mathscr{V}$ is $n-1$ and $\mathscr{V}$ can be specified in the following equivalent form:

$$\mathscr{V} = \{\mathbf{v} \mid \mathbf{Bv} = \mathbf{0}, \, \mathbf{v} \in \mathbb{C}^m\}. \tag{A.1.15}$$

Utilizing (A.1.8), the system of equations defining the voltage space can also be formulated in the following form:

$$\mathbf{v}_K + \mathbf{F}\mathbf{v}_F = \mathbf{0}. \tag{A.1.16}$$

The components of vectors $\mathbf{v} \in \mathscr{V}$ will also be called edge voltages.

From (A.1.16) it can be seen that choosing arbitrarily the edge voltages for the edges of the tree, the edge voltages of the links are uniquely determined.

The relation between the voltage and current spaces is formulated in the following theorem.

Theorem 12 *The voltage space $\mathscr{V}$ and the current space $\mathscr{I}$ are orthogonal complementary subspaces.*

Proof. From the definitions, by applying Theorem 10, we obtain $\mathscr{I} = \mathbb{C}_B$, $\mathscr{V} = \mathbb{C}_Q$. From these the proposition follows by applying Theorem 10. □

In the sequel we will need the following notation. For any vector $\mathbf{a} \in \mathbb{C}^q$ ($q \geq 1$), let $\mathbf{a}^* \in \mathbb{C}^q$ denote a vector that is obtained by replacing all components with their complex conjugates in $\mathbf{a}$.

Definition The directed graph $(\mathscr{N}, \mathscr{A}^D)$, along with an assignment of edge voltages and edge currents, will be called a network, provided that the voltage and current vectors are elements of the voltage space $\mathscr{V}$ and the current space $\mathscr{I}$, respectively.

Definition Let $\mathbf{v} \in \mathscr{V}$ and $\mathbf{i} \in \mathscr{I}$. Then the electric power attached to the edges of the network can be defined as follows.

If $(j,l) \in \mathscr{A}^D$, then the power corresponding to (j,l), denoted by s_{jl}, is the following quantity:

$$s_{jl} = v_{jl} i_{jl}^*. \tag{A.1.17}$$

The vector of powers attached to the edges is denoted by $\mathbf{s}$, and its components will be called edge powers.

The total power S attached to the network is defined as the sum of the edge powers. Therefore, on the basis of (A.1.12) we obtain

$$S = \mathbf{v}^T \mathbf{i}^* = \langle \mathbf{v}, \mathbf{i} \rangle = 0. \tag{A.1.18}$$

Considering the physical content, the orthogonality of $\mathscr{I}$ and $\mathscr{V}$ describes power conservation.

Definition The electrical states of the network $(\mathscr{N}, \mathscr{A}^D)$ are defined as pairs of vectors $(\mathbf{i}, \mathbf{v})$, $\mathbf{i} \in \mathscr{I}$, $\mathbf{v} \in \mathscr{V}$. Therefore, the set of all electrical states is $\mathscr{I} \times \mathscr{V}$.

Generally, in physical networks not every $(\mathbf{i}, \mathbf{v})$, $\mathbf{i} \in \mathscr{I}$, $\mathbf{v} \in \mathscr{V}$ pair is feasible. For this reason, the notion of feasible electrical states is introduced. From a mathematical point of view, this means that an $\mathscr{M} \subseteq \mathscr{I} \times \mathscr{V}$ set is fixed, and an electrical state $(\mathbf{i}, \mathbf{v})$ is called feasible if $(\mathbf{i}, \mathbf{v}) \in \mathscr{M}$ is satisfied.

In the sequel, we discuss two cases concerning the possible choices of $\mathscr{M}$.

Definition The linear admittance transformation of the network $(\mathscr{N}, \mathscr{A}^D)$ is a linear transformation mapping $\mathbb{C}^m$ into itself, that is, mapping the voltage space $\mathscr{V}$ into the current space $\mathscr{I}$. The matrix of this transformation in the natural basis will be denoted by $\mathbf{Y}$.

Then matrix $\mathbf{Y}$ must fulfill the following condition. If $\mathbf{v} \in \mathscr{V}$, then $\mathbf{Yv} = \mathbf{i} \in \mathscr{I}$ must hold.

Let $\mathscr{I}_Y$ denote the image space of the voltage space $\mathscr{V}$ resulting from the transformation, i.e.,

$$\mathscr{I}_Y = \{\mathbf{i} \mid \text{ there exists a } \mathbf{v} \in \mathscr{V}, \text{ such that } \mathbf{i} = \mathbf{Yv} \text{ holds}\}. \tag{A.1.19}$$

Then the requirement concerning matrix $\mathbf{Y}$ can also be formulated as $\mathscr{I}_Y \subseteq \mathscr{I}$.

If an admittance transformation with matrix $\mathbf{Y}$ is attached to the network $(\mathscr{N}, \mathscr{A}^D)$, the set $\mathscr{M}_Y$ of the feasible electrical states is the following subset of $\mathscr{I}_Y \times \mathscr{V}$:

$$\mathscr{M}_Y = \{(\mathbf{i}, \mathbf{v}) \mid \mathbf{i} = \mathbf{Yv}, \mathbf{i} \in \mathscr{I}, \mathbf{v} \in \mathscr{V}\}. \tag{A.1.20}$$

Let us now consider a spanning tree $(\mathscr{N}, \mathscr{F}^D)$, and let us partition matrix $\mathbf{Y}$ according to the link and tree edges (branches) in the following way:

$$\mathbf{Y} = \begin{pmatrix} \mathbf{Y}_{KK}, \mathbf{Y}_{KF} \\ \mathbf{Y}_{FK}, \mathbf{Y}_{FF} \end{pmatrix}. \tag{A.1.21}$$

Then the partitioned form of the admittance transformation will be as follows:

$$\begin{pmatrix} \mathbf{i}_K \\ \mathbf{i}_F \end{pmatrix} = \begin{pmatrix} \mathbf{Y}_{KK}, \mathbf{Y}_{KF} \\ \mathbf{Y}_{FK}, \mathbf{Y}_{FF} \end{pmatrix} \begin{pmatrix} \mathbf{v}_K \\ \mathbf{v}_F \end{pmatrix}. \tag{A.1.22}$$

On the basis of (A.1.16), the voltage space $\mathscr{V}$ can be represented as

$$\mathscr{V} = \left\{ \mathbf{v} \,\middle|\, \mathbf{v} = \begin{pmatrix} \mathbf{v}_K \\ \mathbf{v}_F \end{pmatrix} = \begin{pmatrix} -\mathbf{F} \\ \mathbf{E}_{n-1} \end{pmatrix} \mathbf{v}_F, \ \mathbf{v}_F \in \mathbb{C}^{n-1} \right\}. \tag{A.1.23}$$

Substituting this into relation (A.1.22), we obtain

$$\begin{pmatrix} \mathbf{i}_K \\ \mathbf{i}_F \end{pmatrix} = \begin{pmatrix} \mathbf{Y}_{KK}, \mathbf{Y}_{KF} \\ \mathbf{Y}_{FK}, \mathbf{Y}_{FF} \end{pmatrix} \begin{pmatrix} -\mathbf{F} \\ \mathbf{E}_{n-1} \end{pmatrix} \mathbf{v}_F. \qquad (A.1.24)$$

Thus, we get

$$\mathbf{i}_K = (-\mathbf{Y}_{KK}\mathbf{F} + \mathbf{Y}_{KF})\mathbf{v}_F,$$

$$\mathbf{i}_F = (-\mathbf{Y}_{FK}\mathbf{F} + \mathbf{Y}_{FF})\mathbf{v}_F. \qquad (A.1.25)$$

On the basis of the definition, $\mathbf{Y}$ is an admittance transformation if and only if $\mathbf{i} = \begin{pmatrix} \mathbf{i}_K \\ \mathbf{i}_F \end{pmatrix}$, computed according to (A.1.25), is an element of $\mathscr{I}$ for any $\mathbf{v}_F \in \mathbb{C}^{n-1}$. Utilizing the defining Eq. (A.1.13) of the current space, we get that for any $\mathbf{v}_F \in \mathbb{C}^{n-1}$, the following relation must hold:

$$(\mathbf{F}^T\mathbf{Y}_{KK}\mathbf{F} - \mathbf{F}^T\mathbf{Y}_{KF})\mathbf{v}_F = (\mathbf{Y}_{FK}\mathbf{F} - \mathbf{Y}_{FF})\mathbf{v}_F. \qquad (A.1.26)$$

Consequently, for the submatrices of $\mathbf{Y}$ the following condition holds:

$$\mathbf{F}^T\mathbf{Y}_{KK}\mathbf{F} - \mathbf{F}^T\mathbf{Y}_{KF} = \mathbf{Y}_{FK}\mathbf{F} - \mathbf{Y}_{FF}. \qquad (A.1.27)$$

Further on, we will discuss the case where $\mathbf{Y}_{KF} = \mathbf{0}$, $\mathbf{Y}_{FK} = \mathbf{0}$ holds. Regarding its physical background, this means that only those networks are analyzed that contain a spanning tree such that the electromagnetic connection (mutual admittances) between tree edges and link edges is negligible.

In this case the matrix of the admittance transformation is

$$\mathbf{Y} = \begin{pmatrix} \mathbf{Y}_{KK} & \mathbf{0} \\ \mathbf{0} & -\mathbf{F}^T\mathbf{Y}_{KK}\mathbf{F} \end{pmatrix}. \qquad (A.1.28)$$

Therefore, the admittance transformation takes the following form:

$$\mathbf{i}_K = \mathbf{Y}_{KK}\mathbf{v}_K, \qquad (A.1.29)$$

$$\mathbf{i}_F = -\mathbf{F}^T\mathbf{Y}_{KK}\mathbf{F}\mathbf{v}_F. \qquad (A.1.30)$$

Applying the preceding relations and relation (A.1.16), the subspace $\mathscr{I}_Y$ can be represented as

$$\mathscr{I}_Y = \left\{ \mathbf{i} \,\middle|\, \mathbf{i} = -\begin{pmatrix} \mathbf{Y}_{KK}\mathbf{F} \\ \mathbf{F}^T\mathbf{Y}_{KK}\mathbf{F} \end{pmatrix} \mathbf{v}_F, \ \mathbf{v}_F \in \mathbb{C}^{n-1} \right\}. \qquad (A.1.31)$$

The physical background of relations (A.1.29) and (A.1.30) is as follows. The diagonal elements of matrix $\mathbf{Y}_{KK}$ correspond to the edge admittances of the links, while the rest of the elements in the matrix relate to the mutual admittances between links. (Edge admittances in the physical sense will be discussed in the next section; see also [70]). Considering relation (A.1.16) as well, and in accordance with (A.1.29) and (A.1.30), we conclude that when fixing the voltages of tree edges, for a given $\mathbf{Y}_{KK}$ the currents in the entire network are uniquely determined.

Note that, of course, matrix $\mathbf{Y}$ of the admittance transformation is not the same as the matrix constructed on the basis of the admittances of the passive elements of the network (see [70]), the latter also being denoted by $\mathbf{Y}$ in the engineering literature.

Definition The (linear) impedance transformation of the network $(\mathcal{N}, \mathscr{A}^D)$ is a linear transformation mapping $\mathbb{C}^m$ into itself, which maps the current space $\mathscr{I}$ into the voltage space $\mathscr{V}$. The matrix of the transformation in the natural basis will be denoted by $\mathbf{Z}$. Then matrix $\mathbf{Z}$ must fulfill the following condition. If $\mathbf{i} \in \mathscr{I}$, then $\mathbf{Z}\mathbf{i} = \mathbf{v} \in \mathscr{V}$ holds.

Let $\mathscr{V}_Z$ denote the image space of the current space $\mathscr{I}$ under an impedance transformation, i.e., the following subspace of $\mathbb{C}^m$:

$$\mathscr{V}_Z = \{\mathbf{v} \mid \text{There exists an } \mathbf{i} \in \mathscr{I} \text{ for which } \mathbf{v} = \mathbf{Z}\mathbf{i} \text{ holds}\}. \qquad (\text{A.1.32})$$

Then the requirement concerning $\mathbf{Z}$ can also be given in the following form: $\mathscr{V}_Z \subseteq \mathscr{V}$.

If an impedance transformation with matrix $\mathbf{Z}$ is assigned to the network $(\mathcal{N}, \mathscr{A}^D)$, then the set $\mathscr{M}_Z$ of the feasible electrical states will be the following subset of the set $\mathscr{I} \times \mathscr{V}_Z$:

$$\mathscr{M}_Z = \{(\mathbf{i}, \mathbf{v}) \mid \mathbf{v} = \mathbf{Z}\mathbf{i}, \mathbf{i} \in \mathscr{I}, \mathbf{v} \in \mathscr{V}\}. \qquad (\text{A.1.33})$$

Let us consider a spanning tree $(\mathcal{N}, \mathscr{F}^D)$ and partition matrix $\mathbf{Z}$ accordingly. Proceeding analogously as in the case of the admittance transformation, the following matrix $\mathbf{Z}$ is obtained:

$$\mathbf{Z} = \begin{pmatrix} -\mathbf{F}\mathbf{Z}_{FF}\mathbf{F}^T & \mathbf{0} \\ \mathbf{0} & \mathbf{Z}_{FF} \end{pmatrix}. \qquad (\text{A.1.34})$$

The impedance transformation takes the following form:

$$\mathbf{v}_K = -\mathbf{F}\mathbf{Z}_{FF}\mathbf{F}^T\mathbf{i}_K, \qquad (\text{A.1.35})$$

$$\mathbf{v}_F = \mathbf{Z}_{FF}\mathbf{i}_F. \qquad (\text{A.1.36})$$

From this, applying (A.1.13), we get the following representation of subspace $\mathcal{V}_Z$:

$$\mathcal{V}_Z = \left\{ \mathbf{v} \,\middle|\, \mathbf{v} = \begin{pmatrix} -\mathbf{F}\mathbf{Z}_{FF}\mathbf{F}^T \\ \mathbf{Z}_{FF}\mathbf{F}^T \end{pmatrix} \cdot \mathbf{i}_K, \mathbf{i}_K \in \mathbb{C}^k \right\}. \tag{A.1.37}$$

The physical background of the impedance transformation is as follows.

The diagonal elements of matrix $\mathbf{Z}_{FF}$ correspond to the edge impedances of the tree edges (see the following section or, for example, [70]), while the rest of the elements relate to the mutual impedances among tree edges (e.g., [70]). According to (A.1.35) and (A.1.36) and taking (A.1.13) into account, we conclude that if the currents of the link edges are fixed, then the voltages are uniquely determined in the whole network for a given $\mathbf{Z}_{FF}$.

Of course, matrix $\mathbf{Z}$ of the impedance transformation as defined earlier is not the same as the matrix being built on the basis of the impedances of passive elements, which is also denoted by $\mathbf{Z}$ in the engineering literature.

Both matrix $\mathbf{Y}$ of the admittance transformation and $\mathbf{Z}$ of the impedance transformation can be singular, even in the case where $\mathbf{Y}_{KK}$ and $\mathbf{Z}_{FF}$ are nonsingular matrices. In our case, the invertibility of the transformations restricted to the voltage respectively current spaces is essential.

In the case of the admittance transformation, this means that we consider the transformation defined by the relation $\mathbf{i} = \mathbf{Y}\mathbf{v}$ but restricted to $\mathbf{v} \in \mathcal{V}$ only. This is a linear mapping $\mathcal{V} \to \mathcal{I}_Y$. For ensuring that it is a one-to-one correspondence, the condition $\dim \mathcal{V} = n - 1 = k = \dim \mathcal{I}$ must necessary be fulfilled. It can be seen from (A.1.31) that it is a one-to-one mapping if and only if the columns of the involved matrices are linearly independent. If the inverse transformation exists, it is called a restricted impedance transformation. Its existence means that for any $\mathbf{i} \in \mathcal{I}_Y$ there exists one and only one $\mathbf{v} \in \mathcal{V}$ for which $(\mathbf{i}, \mathbf{v})$ is a feasible electrical state.

Similar statements hold in the case of the impedance transformation, and the restricted impedance transformation is defined in an analogous way.

Considering its physical background, assigning admittance and impedance transformations to the networks can be regarded as a generalization of *Ohm's law* to networks.

Note that our starting assumption on fixing a spanning tree $(\mathcal{N}, \mathcal{F}^D)$ has no conceptual significance, apart from simplifying the presentation. As a result of the theorems proved earlier, if different spanning trees are chosen for the voltage and current spaces, this merely means a change of the basis in those subspaces.

As an illustration, we consider a simple circuit. By this is meant the following network:

$$\mathcal{N} = \{1,2\}, \quad \mathcal{A}^D = \{(1,2), (2,1)\}, \quad n = m = 2, \quad k = 1.$$

The current space is the subspace $\mathcal{I} = \{(i_1, i_2) \mid i_1 = i_2\}$, while the voltage space is the subspace $\mathcal{V} = \{(v_1, v_2) \mid v_1 = -v_2\}$. Both of them are one-dimensional subspaces of $\mathbb{C}^2$. In our case, the specification of the impedance and admittance transformations is equivalent to choosing one of the edges. Let us choose $(2, 1)$, for example, and fix a relation of the form

$$v_{21} = Z_{21} i_{21}, \tag{A.1.38}$$

where $Z_{21} \in \mathbb{C}$ is a constant.

Considering its physical content, (A.1.38) is the usual form of *Ohm's law*. The matrix of the impedance transformation is as follows:

$$\mathbf{Z} = \begin{pmatrix} -Z_{21} & 0 \\ 0 & Z_{21} \end{pmatrix}. \tag{A.1.39}$$

The restricted impedance and admittance transformations exist; they are one another's inverses. Therefore, they provide a one-to-one transformation between $\mathscr{I}$ and $\mathscr{V}$.

Definition The quantity Z_{21} in relation (A.1.38) is called the edge impedance of edge $(2, 1)$, while the quantity $\mathbf{Y}_{21} = \dfrac{1}{Z_{21}}$ is called an edge admittance of the edge.

Finally, let us note that the presented theory can be developed in an analogous way when taking $\mathbb{R}^m$ instead of $\mathbb{C}^m$ as the underlying space.

A.2 Physical Description of the Transmission Network of Electric Power Systems

In this section the construction of a surrogate electric network (in the physical sense of the word) for a transmission network is outlined. Under the assumptions outlined in this section, the voltage-current conditions of this surrogate network reflect the conditions of the actual transmission network with reasonably good accuracy.

After an introductory discussion concerning harmonic alternating currents, a simplified description of the transmission network follows. We mostly concentrate on those components of the network that significantly influence the voltage-current conditions and, therefore, the power conditions. Then we discuss the (in the physical sense) substitute electric networks of the main components of a transmission network. Finally, the structure of the substitute overall electric network is presented.

In the course of the construction of the model of the transmission network, harmonic alternating currents are dealt with. To summarize the related physical notions and basic relations, we consider the simple circuit displayed in Fig. A.6.

In this figure G is a voltage source and A represents one of the following ideal network elements: ohmic resistance, inductor (coil), capacitor.

If the voltage source produces a harmonic alternating voltage $v(t)$ having angular frequency ω, then, if A is replaced by any of the previously listed ideal network

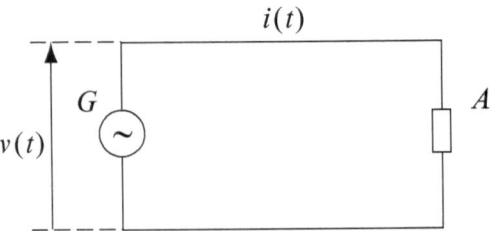

Fig. A.6 Simple circuit

elements, the current $i(t)$ flowing in the circuit will be a harmonic alternating current with the same angular frequency ω. This means that if the voltage is

$$v(t) = v_{max}\cos(\omega t + \varphi), \qquad (\text{A.2.1})$$

then

$$i(t) = i_{max}\cos(\omega t + \psi) \qquad (\text{A.2.2})$$

holds for the current flowing in the circuit, where v_{max} and i_{max} are the amplitudes and φ and ψ are the phase-angles for the voltage and current, respectively.

If A represents ohmic resistance with resistance value R, then, with respect to voltage and current,

$$v(t) = Ri(t) \qquad (\text{A.2.3})$$

holds at any time t.

However, if A is an inductor with inductance L, the differential equation

$$v(t) = L\frac{di(t)}{dt} \qquad (\text{A.2.4})$$

describes the relation between voltage and current.

The situation is similar in the case of a capacitor of capacitance C where the differential equation takes the following form:

$$i(t) = C\frac{dv(t)}{dt}. \qquad (\text{A.2.5})$$

Next, the complex formalism concerning alternating currents is outlined. For the rest of the book let j denote the complex imaginary unit.

Let us introduce the following notations:

$$V = \frac{v_{max}}{\sqrt{2}} e^{j\varphi}, \tag{A.2.6}$$

$$I = \frac{i_{max}}{\sqrt{2}} e^{j\psi}. \tag{A.2.7}$$

V and I, as defined by (A.2.6) and (A.2.7), are called the phase representations of voltage and current, respectively. The role of the factor $\frac{1}{\sqrt{2}}$ will become clear in the subsequent discussion on complex power.

Substituting (A.2.1) and (A.2.2) into relations (A.2.3)–(A.2.5), it is easy to see that (A.2.3)–(A.2.5) are respectively formally equivalent to

$$V = RI, \tag{A.2.8}$$

$$V = j\omega LI, \tag{A.2.9}$$

$$V = \frac{1}{j\omega C} I. \tag{A.2.10}$$

The new formulas can be regarded as a complex representation of *Ohm's law*. The complex quantities (acting as the coefficient of I in the relations) are called the impedances of the respective ideal circuit elements, while their reciprocals are called admittances. Let Z denote the impedance defined in this way. Its value for the various circuit elements is as follows:

$$Z = R \qquad \text{for ohmic resistance,} \tag{A.2.11}$$

$$Z = j\omega L \quad \text{for inductor,} \tag{A.2.12}$$

$$Z = \frac{1}{j\omega C} \quad \text{for capacitor.} \tag{A.2.13}$$

In the sequel R will be called resistive reactance, ωL inductive reactance, and $\frac{1}{\omega C}$ capacitive reactance.

If we build an electric network in a physical sense on the basis of the ideal network elements and voltage source discussed so far, then for the instantaneous values of current and voltage *Kirchhoff's laws* hold. Assuming that the angular frequency ω is the same throughout the network, it is easy to see that *Kirchhoff's current law* concerning the nodes can be formulated in an equivalent way by replacing the instantaneous values of currents with their phase representation. The same is true concerning *Kirchhoff's voltage law* on circuits and the phase representation of the voltages.

In the sequel we assume that for all current and voltage sources the angular frequency ω is the same.

It can be seen that the model described in Sect. A.1 is quite suitable for the mathematical description of a physical network built up from ideal circuit elements by assigning the different network elements to the edges of the graph and thinking in terms of phase representation. It is easy to show that if network elements are connected in series or in parallel, the new compound element (due to *Ohm's law* formulated according to the complex description) can be substituted by a single impedance that can be computed on the basis of the impedances of the components, in an analogous way to the case of direct currents. Therefore, to reduce the dimension of the mathematical model network, we associate to each of the compound elements discussed earlier a single edge in the graph.

Regarding the circuit in Fig. A.6, the corresponding mathematical model will be a simple circuit as defined and discussed in Sect. A.1. The impedances of the network elements correspond to impedances assigned to the edges of the graph.

Let us recall the concept of electric power for harmonic alternating currents. Consider again the circuit in Fig. A.6. Here the instantaneous power, denoted by $p(t)$, is defined as follows:

$$p(t) = v(t)i(t) = v_{max}i_{max}\cos(\omega t + \varphi)\cos(\omega t + \psi). \tag{A.2.14}$$

Following some elementary computations this can be written as

$$p(t) = \frac{1}{2}v_{max}i_{max}\cos(\varphi - \psi)[1 + \cos 2(\omega t + \psi)]$$
$$- \frac{1}{2}v_{max}i_{max}\sin(\varphi - \psi)\sin 2(\omega t + \psi). \tag{A.2.15}$$

Let us introduce the notations as specified below:

$$P = \frac{1}{2}v_{max}i_{max}\cos(\varphi - \psi), \tag{A.2.16}$$

$$Q = \frac{1}{2}v_{max}i_{max}\sin(\varphi - \psi). \tag{A.2.17}$$

With these notations, the equation for the instantaneous power takes the form

$$p(t) = P(1 + \cos 2(\omega t + \psi)) - Q\sin 2(\omega t + \psi), \tag{A.2.18}$$

where P represents active power, which can be easily shown to be the mean value of instantaneous power over one period. The first term in (A.2.18) can be interpreted as a power flow of varying magnitude. Its direction depends on the sign of P, and its values vary in a harmonic way around the mean value P by an angular frequency of 2ω ($P \geq 0$ corresponds to consumption).

Q represents reactive power. According to the second term in relation (A.2.18), it is the peak value of a harmonic oscillation of a 2ω angular frequency and of mean value 0. Physically, the reactive power represents a continuous exchange of energy of mean value 0 between the voltage source and the impedance. More precisely, this exchange of energy takes place between the source and the electromagnetic field of the impedance. It can easily be derived from the preceding definition of Q and Eqs. (A.2.4) and (A.2.5) that inductive impedance consumes reactive power ($Q \geq 0$), while capacitive impedance produces reactive power ($Q \leq 0$).

Complex power S is introduced with the following definition by applying the phase representation of voltage (A.2.6) and current (A.2.7):

$$S = VI^*. \tag{A.2.19}$$

Substituting the phase representations, for complex power we get the following formulation:

$$S = \frac{1}{2} v_{max} i_{max} e^{j(\varphi - \psi)}. \tag{A.2.20}$$

Therefore, S can be expressed in terms of active and reactive power in the following way:

$$S = P + jQ. \tag{A.2.21}$$

The quantity $|S| = \sqrt{P^2 + Q^2}$ is called apparent power.

In this book the physical quantities are considered as being measured according to the following units:

Impedances (resistive, inductive, capacitive, reactance)	Ω
Admittances (all three)	$S = \dfrac{1}{\Omega}$
Voltage	kV
Current	kA
Active power	MW
Reactive power	Mvar
Apparent power	MVA.

Now we will discuss the transmission network of electric power systems.

In general, transmission networks are many-phased. Their characteristic (and herein analyzed) mode of operation is a symmetric balanced state, meaning that in the individual phases the size of the current is the same and the phase angles are shifted by constant values with respect to each other. In this case the determination of the electrical state of the network can be carried out by a single-phase substitution; based on the results obtained in this way, the actual state can be reconstructed. Therefore, in the sequel it is sufficient to consider single-phase networks only. In the networks considered, in steady states the frequency of the harmonic alternating current has the same value across the network. Consequently, the complex phase representation discussed previously can be applied in the construction of the model.

In an electric power system the power generated in power plants is conveyed to consumers via an electric transmission network. In our model the basic network is considered only, i.e., transmission lines of 750 kV, 400 kV, and 220 kV and that part of the main distribution network that is 120 kV, selected according to power systems engineering criteria, which plays a substantial role in power distribution.

From an engineering point of view, power transmission is carried out by transmission lines and cables. Subnetworks, corresponding to different voltage levels, are connected to one another through transformers.

The nodes of the network are power plant busbars and consumer substations. The term *busbar* originates from the fact that the branches of the network, the generators and transformers of the power plants, and the transformers connecting the consumer distribution network of lower voltage to the transmission network are connected by busbars made of copper or aluminum at the various network substations.

Next those main devices are enlisted whose ensemble results in power injection respectively consumption at the nodes.

- *Power plant synchronous generators* generate active power and, depending on their excitation control, may generate or consume reactive power. In our model the overall power of all generators (possibly located in various power plants) connected to the same node is considered.
- *Consumers* (load) includes all consumers connected to a node through the distribution network of a lower voltage level. In our model the overall power demand at the nodes is taken into account.

The following devices are used to inject or consume reactive power.

- A *synchronous compensator* is a synchronous engine that does not inject active power at a node. Depending on the excitation of the rotor, it can either consume or generate reactive power within the range delimited by its domain of control.
- A *shunt reactor* is an inductive device that consumes reactive power.
- A *shunt capacitor*, due to its capacitive characteristic, generates reactive power. Several shunt capacitors can be connected to one node. Then the capacity can be changed within the given limits by discrete jumps.

Next, the equivalent circuits of branches and transformers of a transmission network are discussed. These are simple electric circuits consisting of a few edges and ideal network elements. By an appropriate selection of the involved impedance values they reflect the voltage-current states of the corresponding element of the transmission network, in a steady state of the network, with reasonable accuracy from an engineering viewpoint. Naturally, the equivalent circuit can only provide a true picture of the voltage and current of the element of the transmission network in that element's domain of stable operation.

As a substitute for transmission lines, cables, and transformers, the so-called π circuit can be utilized. This consists of three nodes and three branches, as illustrated in Fig. A.7.

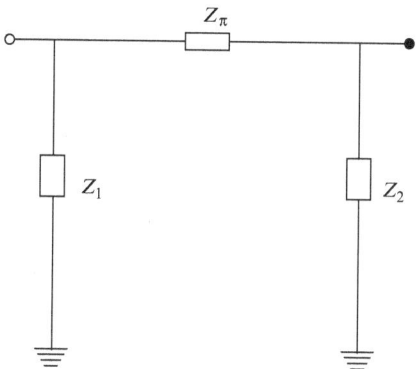

Fig. A.7 A π circuit

One of the three nodes is the grounded node, while the other two correspond to busbars of the transmission network. The impedance values Z_π, Z_1, and Z_2 are specified on the basis of the engineering parameters of the transmission lines, cables, and transformers [17]. In all three cases Z_π takes the form $Z_\pi = R + j\omega L$, where R represents the ohmic resistance of the relevant element and L its inductance. An important characteristic of the transmission network is that for all branches the inductive reactance ωL is significantly higher than the ohmic resistance R. In the case of transformers these quantities differ by one order of magnitude.

In the case of transmission lines and cables, $Z_1 = Z_2 = \dfrac{2}{j\omega C}$ holds, where C represents the capacitance of transmission lines and cables. In the case of transmission lines the value of ωC is by approximately six orders of magnitudes smaller than the value of ωL. The value of capacity assigned to cables is significantly higher than the values of capacity attached to transmission lines.

In the case of the transformers the situation is more complicated because some of them can be regulated having a variable turn ratio; these are so-called tap-changing transformers. In the case of non-tap-changing transformers and those of the tap-changing ones for which the turn ratio is at a nominal value, the equivalent circuit shown in Fig. A.7 simplifies because the vertical parts of π can be neglected. Therefore, they can be represented by a single branch with an appropriately chosen impedance value Z_π.

In the case of those tap-changing transformers for which the turn ratio is not at a nominal value, the values of the edge impedances Z_π, Z_1, and Z_2 displayed in Fig. A.7 can be determined on the basis of the transformer specifications and transformer tap adjustments [17].

In the sequel, for the sake of simplicity of presentation we assume that for all of the tap-changing transformers the turn ratio is at a nominal value; thus, all of them can be represented by single branches with appropriately chosen Z_π impedance values.

Based on the foregoing discussions, we will build a physical model of a transmission network, that is, we will construct an electric network whose voltage-current states reflect the voltage-current states of the transmission network with reasonable accuracy.

The nodes of the equivalent electric network correspond to busbars. Let their number be denoted by N. It is also assumed that the nodes are numbered. As an $(N + 1)$th node we introduce the grounded node.

The branches of the network are constructed as follows.

Let us assume that the ith and lth busbars are connected by a transmission line or cable. In the equivalent circuit shown in Fig. A.7, resistance, coefficient of inductance, and capacitance are denoted by R_{il}, L_{il}, and C_{il}, respectively. In the equivalent electric network, we introduce a branch connecting the ith node with the lth node that corresponds to the upper horizontal part of π with an edge impedance value of $R_{il} + j\omega L_{il}$. The vertical parts of π connect the ith and lth nodes with the grounded node; an impedance value of $\dfrac{2}{j\omega C_{il}}$ is assigned to both branches.

If the ith and lth busbars are connected by a transformer and R_{il}, L_{il} ohmic resistance and coefficient of inductance has been assigned to it, in the equivalent electric network the transformer is represented by an edge connecting the ith and lth nodes with an edge impedance value of $R_{il} + j\omega L_{il}$.

If the ith and lth nodes are connected by several transmission lines or cables, they can be reduced in the equivalent network to the preceding three branches by a substitution valid for impedances connected in parallel. For the sake of simplicity, let R_{il}, L_{il}, C_{il} denote the resultant resistance, coefficient of inductance, and capacitance, respectively.

Let M denote the number of those branches in the previously described electric network whose two endpoints correspond to busbars. For the sake of simplicity of reference, henceforth, these types of branches will be called γ-type branches.

In the electric network achieved as a result of the procedure outlined earlier, apart from the grounded node, the same number of capacitive branches (grounded at the other ends) are connected to each node as there are γ-type branches connected to that node that represent transmission lines or cables. At each node these capacitive branches are connected in parallel and can be replaced by a single branch, with the capacitance of this branch being the sum of the capacitances of those branches. If only γ-type branches corresponding to transformers are connected to a node, then no such branch is introduced. Let N_C denote the number of those nodes to which we have assigned a branch in the previously described way, connecting that node to the grounded node, and let us assume that the serial numbers of the nodes start with these nodes. Therefore, to each of the first N_C nodes there is connected a capacitive branch of the aforementioned type. These capacitive branches will be called β-type branches. Let the capacitance of these branches be denoted by C_i, $i = 1, \ldots, N_C$.

As discussed previously, as a result of the operation of several components there is a concentrated power injection or demand at busbars of the transmission network. To take this into account, for each of the nodes corresponding to busbars the electric network is supplemented by an additional branch connecting that node to the grounded node. The interpretation is that the power injection or demand concentrated at the nodes arises in these branches. There are N branches of this type that will be called α-type branches. However, no edge impedance values can be assigned to these new branches.

In the transmission network the electromagnetic interaction between the various transmission lines, cables, and transformers can be neglected. Therefore, the same thing is assumed with respect to the β and γ-type branches of the electric network. It is also assumed that there is no electromagnetic interaction between the α and the β respectively γ-type branches. This means that with respect to the β and γ-type branches the voltage-current relation is the same as if the relevant branches were located where circuit element A is located in the circuit displayed in Fig. A.6.

Summarizing, in an electric network, the number of nodes is $N + 1$, and the $(N + 1)$th node is the grounded node.

α-type branches: branches connecting nodes with serial number $1, \ldots, N$ to the grounded node. They have no edge impedance value assigned to them. The number of such edges in the network is N.

β-type branches: branches connecting nodes with serial numbers $1, \ldots, N_C$ to the grounded node of serial number $N + 1$. To the ith branch of this type is assigned the edge impedance $\dfrac{1}{j\omega C_i}$. The number of β type branches is N_C.

γ-type branches: branches whose serial numbers of the two endpoints are $\leq N$. The set of these branches can be described mathematically as a subset of the set of all pairs of numbers from the set $\{1, \ldots, N\}$. If this pair of numbers is $\{i, l\}$, the edge impedance value assigned to the corresponding branch is $R_{il} + j\omega L_{il}$. The number of γ-type branches is M.

Additional significant, though not yet discussed, characteristics of a network are the following. Apart from grounded nodes, an undirected graph corresponding to nodes and γ-type branches is connected. Furthermore, the network is "looped," i.e., the corresponding graph contains several circuits. Therefore, it can be assumed that $M > N$ holds.

A.3   Mathematical Model of the Transmission Network of Electric Power Systems

Our main objective in this section is, starting from the physical model described in Sect. A.2, to set up a mathematical model of a transmission network by applying the general theory presented in Sect. A.1.

First a network model is built that serves as the basis of the mathematical model of our transmission network. Then its voltage and current spaces are analyzed and the notion of transmission losses is introduced and discussed. Starting from the system of equations resulting from an analysis of an admittance transformation, the notion of a reference node is introduced. In the course of our discussion of these topics, we will use the notation introduced in both of the previous sections. In cases where this may lead to conflicting interpretations, we will make clear the sense in which the notation is being used.

Finally, we formulate the main problem of load flow analysis for transmission networks. For this we introduce some new notations, adapted to the standard notation in the engineering literature.

Relying on the electric network model constructed in the previous section, the mathematical model of the transmission network will be the following specified special case of the general model presented in Sect. A.1.

The network is represented by the following graph $(\mathcal{N}, \mathcal{A}^D)$:

$\mathcal{N}$: the set of graph nodes consisting of $n = N + 1$ elements, where the node corresponding to the grounded node of the electric network has the serial number $N + 1$;

$\mathcal{A}^D$: the set of graph edges having $m = M + N + N_C$ elements partitioned according to the α-, β-, and γ-type branches of the electric network.

Let

$$\mathcal{A}^D = \mathcal{A}^D_\alpha \cup \mathcal{A}^D_\beta \cup \mathcal{A}^D_\gamma, \tag{A.3.1}$$

where the sets of edges $\mathcal{A}^D_\alpha$, $\mathcal{A}^D_\beta$, $\mathcal{A}^D_\gamma$ are defined as follows.

$\mathcal{A}^D_\alpha$: these edges correspond to the α-type branches of the electric network. The definition of this set is as follows:

$$\mathcal{A}^D_\alpha = \{(1, N+1), (2, N+1), \ldots, (N, N+1)\}. \tag{A.3.2}$$

$\mathcal{A}^D_\beta$: the set of edges corresponding to the β-type branches of the electric network, defined as

$$\mathcal{A}^D_\beta = \{(N+1, 1)(N+1, 2), \ldots, (N+1, N_C)\}. \tag{A.3.3}$$

$\mathcal{A}^D_\gamma$: the set of edges corresponding to the γ-type branches of the electric network. This set contains M elements and they connect nodes of the same serial numbers as the corresponding γ-type branches in the electric network. The orientation of these edges is irrelevant henceforth; nevertheless, their orientation is considered as being fixed.

The edges of graph $(\mathcal{N}, \mathcal{A}^D)$ have the following serial numbers: the serial numbers start with edges belonging to $\mathcal{A}^D_\alpha$ in an order that is fixed according to (A.3.2). Then the edges of $\mathcal{A}^D_\beta$ follow in an order as defined in (A.3.3), and, finally, the elements of $\mathcal{A}^D_\gamma$ are numbered in an arbitrary order.

Let $\mathcal{N}_\gamma$ denote a subset of the set of nodes without a grounded node, i.e., $\mathcal{N}_\gamma = \{1, \ldots, N\}$.

The graph $(\mathcal{N}, \mathcal{A}^D)$ is obviously connected. On the basis of the comment at the end of the previous section, we assume that the subgraph $(\mathcal{N}_\gamma, \mathcal{A}_\gamma^D)$ is also connected.

The spanning tree is fixed with the prescription $\mathcal{F}^D = \mathcal{A}_\alpha^D$, i.e., the subgraph $(\mathcal{N}, \mathcal{A}_\alpha^D)$ is chosen as the spanning tree. The tree has $n - 1 = N$ edges. The subgraph $(\mathcal{N}, \mathcal{A}_\alpha)$ forms a star of the graph $(\mathcal{N}, \mathcal{A})$, in the graph-theoretical sense. The set $\mathcal{A}_\alpha$ consists of all those edges that connect node $N + 1$ to the remaining nodes.

The set of link edges, corresponding to the previously mentioned choice of the spanning tree, is the set $\mathcal{A}_\beta^D \cup \mathcal{A}_\gamma^D$. The number of links is $k = m - n + 1 = M + N_C$. The graph $(\mathcal{N}, \mathcal{A}_\beta^D \cup \mathcal{A}_\gamma^D)$ corresponding to links is a connected subgraph of the graph $(\mathcal{N}, \mathcal{A}^D)$.

In this section (in contrast to Sect. A.1) let $\hat{\mathbf{A}}$ denote the node-edge incidence matrix of the subgraph $(\mathcal{N}, \mathcal{A}_\beta^D \cup \mathcal{A}_\gamma^D)$, while $\mathbf{A}$ denotes the reduced node-edge incidence matrix that is formed by deleting the $(N + 1)$th row in $\hat{\mathbf{A}}$.

It can easily be seen that the basic cut sets corresponding to the tree become the node cut sets of nodes with serial numbers $1, \ldots, N$ with respect to the graph $(\mathcal{N}, \mathcal{A}^D)$. That is, the ith cut set consists of all edges from $\mathcal{A}^D$, which are connected to the ith node, $i = 1, \ldots, N$. Therefore, on the basis of (A.1.4), the reduced basic cut-set incidence matrix takes the following form:

$$\mathbf{Q} = (\mathbf{A}, \mathbf{E}_N). \tag{A.3.4}$$

From this and from (A.1.8) it follows that a reduced circuit-node incidence matrix can be specified in the following form by the substitution $\mathbf{F} = -\mathbf{A}^T$:

$$\mathbf{B} = (\mathbf{E}_{M+N_C}, -\mathbf{A}^T). \tag{A.3.5}$$

Consequently, the current space $\mathcal{I}$ is a $k = M + N_C$-dimensional subspace of the space $\mathbb{C}^{M+N+N_C}$. Its defining system of equations on the basis of (A.3.4) and (A.1.13) is

$$\mathbf{A}\mathbf{i}_K + \mathbf{i}_F = \mathbf{0}. \tag{A.3.6}$$

It can be seen from this system of equations that if the current vector $\mathbf{i}_K$ is prescribed for the links, i.e., for the elements of $\mathcal{A}_\beta^D \cup \mathcal{A}_\gamma^D$, then vector $\mathbf{i}_F$ of the currents for the tree edges is uniquely determined.

Relation (A.3.6) can also be interpreted as follows. If we restrict ourselves to the network $(\mathcal{N}, \mathcal{A}_\beta^D \cup \mathcal{A}_\gamma^D)$, then we can argue that at its nodes (apart from the grounded node) currents are injected, and the injected currents are the components of $\mathbf{i}_F$. Since $\mathbf{A}$ is a reduced node-edge incidence matrix of this network, in this interpretation (A.3.6) describes a relation between currents injected at the nodes and currents flowing at the edges of the network.

The voltage space $\mathscr{V}$ is an $n - 1 = N$-dimensional subspace of the space $\mathbb{C}^{M+N+N_C}$ whose defining system of equations is based on (A.3.5) and (A.1.16), as follows:

$$\mathbf{v}_K - \mathbf{A}^T \mathbf{v}_F = \mathbf{0}. \tag{A.3.7}$$

It is evident from this relation that if the edge voltage vector $\mathbf{v}_F$ is prescribed arbitrarily for the edges of the tree, i.e., for the elements of $\mathscr{A}_\alpha^D$, then the edge voltages of the links are uniquely determined. With respect to the links in sets $\mathscr{A}_\beta^D$ respectively $\mathscr{A}_\gamma^D$, this has the following meaning: if the link belongs to the set $\mathscr{A}_\beta^D$, then its form is $(N + 1, i)$, where $1 \leq i \leq N_C$ holds. In this case the basic circuit generated by the edge is formed by the edges $(N + 1, i) \in \mathscr{A}_\beta^D$ and $(i, N + 1) \in \mathscr{A}_\alpha^D$. Taking into account the numbering of the edges of the graph as specified previously, we get

$$v_{N+i} = v_i, \qquad i = 1, \dots, N_C. \tag{A.3.8}$$

If $(i, l) \in \mathscr{A}_\gamma^D$ is fulfilled for a link (i, l), then the basic circuit consists of the edges $(i, l) \in \mathscr{A}_\gamma^D$, $(l, N + 1) \in \mathscr{A}_\alpha^D$, $(i, N + 1) \in \mathscr{A}_\alpha^D$. Therefore, on the basis of (A.3.7), the following relation holds:

$$v_{il} = v_i - v_l, \qquad (i, l) \in \mathscr{A}_\gamma^D. \tag{A.3.9}$$

The results can be interpreted in the following way. Let us assign arbitrary $u_i \in \mathbb{C}$ $(i = 1, \dots, N)$ values to the first N nodes of the graph $(\mathscr{N}, \mathscr{A}^D)$, and let $u_{N+1} = 0$. Furthermore, let the vector $\mathbf{v} \in \mathbb{C}^{M+N+N_C}$ be specified as follows.

If $(i, l) \in \mathscr{A}_\beta^D \cup \mathscr{A}_\gamma^D$, then $v_{il} = u_i - u_l$, and $v_i = u_i$, $i = 1, \dots, N$, with respect to the tree edges.

Relations (A.3.8) and (A.3.9) imply that this vector is an element of the voltage space, i.e., $\mathbf{v} \in \mathscr{V}$ holds. Furthermore, it is clear that all the points of the voltage space can be obtained in this way. This form of representation of the points of the voltage space is known in the literature as the *method of node potentials* [70].

The numbers u_i, $i = 1, \dots, N + 1$, attached to the nodes in the foregoing way are called *node potentials* and the vector $\mathbf{u} \in \mathbb{C}^N$, formed by the the first N of these numbers (i.e., disregarding the node that corresponds to the grounded node), is called a *vector of node potentials*. Utilizing this vector, (A.3.7) can be written as follows:

$$\mathbf{v}_K = \mathbf{A}^T \mathbf{u}. \tag{A.3.10}$$

The result can be interpreted by restricting ourselves to the subgraph $(\mathscr{N}, \mathscr{A}_\beta^D \cup \mathscr{A}_\gamma^D)$: the set of nodes of the subgraph is identical with the set of nodes of the original graph; therefore, the potentials can be related to the subgraph as well.

Then, (A.3.10) provides the derivation of the edge voltages of the subgraph from the node potentials as the usual potential difference.

In accordance with the discussion in Sect. A.2, in the case of β and γ type branches that correspond to the links of the mathematical model, *Ohm's law* holds for each branch. That is, the ratio between the complex edge currents and edge voltages is constant, independently of the complex currents and voltages of the remaining branches. In the mathematical model, this means that the relation

$$\mathbf{i}_K = \mathbf{Y}_{KK}\mathbf{v}_K \tag{A.3.11}$$

holds, where $\mathbf{Y}_{KK}$ is a diagonal matrix whose diagonal elements are the edge admittances of the links. Therefore, in accordance with Sect. A.1, there exists an admittance transformation whose matrix, on the basis of (A.1.28) and (A.3.5), is

$$\mathbf{Y} = \begin{pmatrix} \mathbf{Y}_{KK} & 0 \\ 0 & -\mathbf{A}\mathbf{Y}_{KK}\mathbf{A}^T \end{pmatrix}, \tag{A.3.12}$$

where, in the usual way, the matrix is partitioned by link and tree edges. For the tree edges, the admittance transformation provides the following relation:

$$\mathbf{i}_F = -\mathbf{A}\mathbf{Y}_{KK}\mathbf{A}^T\mathbf{v}_F. \tag{A.3.13}$$

We introduce the notation

$$\hat{\mathbf{Y}} = \mathbf{A}\mathbf{Y}_{KK}\mathbf{A}^T. \tag{A.3.14}$$

Then (A.3.13) takes the following form:

$$\mathbf{i}_F = -\hat{\mathbf{Y}}\mathbf{v}_F. \tag{A.3.15}$$

Matrix $\hat{\mathbf{Y}}$ depends solely on the structure of the graph $(\mathcal{N}, \mathcal{A}_\beta^D \cup \mathcal{A}_\gamma^D)$ and on values of the edge admittances of this graph's edges. Restricting our consideration to $(\mathcal{N}, \mathcal{A}_\beta^D \cup \mathcal{A}_\gamma^D)$, in the literature the matrix $\hat{\mathbf{A}}\mathbf{Y}_{KK}\hat{\mathbf{A}}^T$ is called a node admittance matrix (see [70], for example). Matrix $\hat{\mathbf{Y}}$ derives from it by deletion of the row and column that correspond to the grounded node.

If the network $(\mathcal{N}, \mathcal{A}_\beta^D \cup \mathcal{A}_\gamma^D)$ is considered as a linear and passive $(N+1)$-pole (see the definition of this in [70]), then on the basis of (A.3.15), $-\hat{\mathbf{Y}}$ is a so-called input admittance matrix [70].

From relation (A.1.31) it can be seen that the admittance transformation maps the voltage space $\mathcal{V}$ into the following subspace $\mathcal{I}_Y$ of the current space $\mathcal{I}$:

$$\mathcal{I}_Y = \left\{ \mathbf{i} \mid \mathbf{i} = \begin{pmatrix} \mathbf{Y}_{KK}\mathbf{A}^T \\ -\mathbf{A}\mathbf{Y}_{KK}\mathbf{A}^T \end{pmatrix} \mathbf{v}_F, \quad \mathbf{v}_F \in \mathbb{C}^N \right\}. \tag{A.3.16}$$

If the connected nature of the graph $(\mathcal{N}, \mathcal{A}_\beta^D \cup \mathcal{A}_\gamma^D)$ is utilized, it follows from Theorem 8 that the rows of $\mathbf{A}$ are linearly independent. Therefore, since $\mathbf{Y}_{KK}$ is nonsingular, it is easy to show that the rank of the matrix in relation (A.3.16) is N, and its columns are linearly independent.

From this it can be seen that the subspace $\mathcal{I}_Y$ is N-dimensional and that the admittance transformation restricted to voltage space $\mathcal{V}$ establishes a one-to-one correspondence between $\mathcal{V}$ and $\mathcal{I}_Y$. According to the terminology introduced in Sect. A.1, this means that the restricted impedance transformation exists.

Since in accordance with the discussion in Sect. A.2 $k = M + N_C > N$ holds for the dimension of the current space $\mathcal{I}$, $\mathcal{I}_Y$ is a proper subspace of $\mathcal{I}$. Therefore, in the set of feasible electrical states any $\mathbf{v} \in \mathcal{V}$ occurs coupled with an $\mathbf{i} \in \mathcal{I}$, but the converse is not true.

The orthogonality relation (A.1.18), representing the power balance, can be written alternatively in the form

$$\mathbf{v}_K^T \mathbf{i}_K^* + \mathbf{v}_F^T \mathbf{i}_F^* = 0. \tag{A.3.17}$$

According to Sect. A.1, this relation is implied by the definitions of current space and voltage space (i.e., by *Kirchhoff's laws*). A further derivation is provided in this way from relations (A.3.7), (A.3.11), and (A.3.13). Taking the complex conjugates on both sides of relation (A.3.13) and multiplying by vector $\mathbf{v}_F^T$ from the left, the following relation results:

$$\mathbf{v}_F^T \mathbf{A} \mathbf{Y}_{KK}^* \mathbf{A}^T \mathbf{v}_F^* + \mathbf{v}_F^T \mathbf{i}_F^* = 0. \tag{A.3.18}$$

From this, utilizing (A.3.7) and (A.3.11), (A.3.17) can directly be derived.

Relation (A.3.18) can also be written in the form

$$-\mathbf{v}_F \mathbf{i}_F^T = \mathbf{v}_K^T \mathbf{i}_K^*. \tag{A.3.19}$$

Definition The quantity $S^v = \mathbf{v}_K^T \mathbf{i}_K^*$ is called a network loss.

Regarding its physical background, S^v represents the overall complex power loss in a transmission network. Equation (A.3.19) can be regarded as a power balance equation if the quantity on the left-hand side represents the power injected into the nodes of the network $(\mathcal{N}, \mathcal{A}_\beta^D \cup \mathcal{A}_\gamma^D)$, whereas S^v represents the loss at the branches. The minus sign on the left-hand side is due to the fact that the tree branches heading for $\mathcal{A}_\alpha$ are directed toward the grounded node.

Utilizing the node potential vector $\mathbf{u}$, the following power-loss formulas can be easily derived:

$$S^v = -\mathbf{u}^T \mathbf{i}_F^*, \tag{A.3.20}$$

$$S^v = \mathbf{u}^T \hat{\mathbf{Y}}^* \mathbf{u}^*, \tag{A.3.21}$$

$$S^v = \mathbf{u}^T \mathbf{A} \mathbf{Y}_{KK}^* \mathbf{A}^T \mathbf{u}^*. \tag{A.3.22}$$

Next, let us consider again relation (A.3.13). By the substitution $\mathbf{v}_F = \mathbf{u}$ we introduce node potentials, and (A.3.13) assumes the form

$$\mathbf{AY}_{KK}\mathbf{A}^T\mathbf{u} = -\mathbf{i}_F. \tag{A.3.23}$$

The relation obtained in this way will be considered a system of equations for the unknown components of $\mathbf{u}$, regarding $\mathbf{i}_F$ as given. Then (A.3.23) describes the potential distribution on the nodes of the network $(\mathcal{N}, \mathscr{A}_\beta^D \cup \mathscr{A}_\gamma^D)$ for given current injections at the nodes. In the sequel we will analyze the coefficient matrix $\hat{\mathbf{Y}} = \mathbf{AY}_{KK}\mathbf{A}^T$ of size $N \times N$ of the system of equations (A.3.23) in a detailed manner.

The reduced node-edge incidence matrix of the connected graph $(\mathcal{N}, \mathscr{A}_\beta^D \cup \mathscr{A}_\gamma^D)$ has the following structure:

$$\mathbf{A} = (\mathbf{E}_c, \mathbf{A}_\gamma), \tag{A.3.24}$$

where $\mathbf{E}_c = -\begin{pmatrix} \mathbf{E}_{N_c} \\ \mathbf{0} \end{pmatrix}$, $\mathbf{E}_{N_c}$ is a unit matrix of size $N_c \times N_c$, and $\mathbf{A}_\gamma$ is the node-edge incidence matrix of the graph $(\mathcal{N}_\gamma, \mathscr{A}_\gamma^D)$. According to Theorem 8, the row vectors of matrix $\mathbf{A}$ are linearly independent.

Let us partition matrix $\mathbf{Y}_{KK}$ according to the β and γ-type edges:

$$\mathbf{Y}_{KK} = \begin{pmatrix} \mathbf{Y}_\beta & \mathbf{0} \\ \mathbf{0} & \mathbf{Y}_\gamma \end{pmatrix}, \tag{A.3.25}$$

where $\mathbf{Y}_\beta$ and $\mathbf{Y}_\gamma$ are diagonal matrices of the appropriate size.

Consequently, on the basis of (A.3.14), the coefficient matrix takes the following form:

$$\hat{\mathbf{Y}} = \begin{pmatrix} \mathbf{Y}_\beta & \mathbf{0} \\ \mathbf{0} & \mathbf{0} \end{pmatrix} + \mathbf{A}_\gamma \mathbf{Y}_\gamma \mathbf{A}_\gamma^T. \tag{A.3.26}$$

The following notations are introduced:

$$\mathbf{Y}_\beta = j\mathbf{Y}_\beta^K, \qquad \mathbf{Y}_\gamma = \mathbf{Y}_\gamma^v + j\mathbf{Y}_\gamma^K, \tag{A.3.27}$$

where $\mathbf{Y}_\beta^K$, $\mathbf{Y}_\gamma^v$, $\mathbf{Y}_\gamma^K$ are real matrices. Then, by straightforward computations, we get that matrix $\tilde{\mathbf{Y}}$ of the real system of equations that corresponds to the complex system of Eqs. (A.3.23) takes the following form:

$$\tilde{\mathbf{Y}} = \begin{pmatrix} \mathbf{0} & -\tilde{\mathbf{Y}}_\beta \\ \tilde{\mathbf{Y}}_\beta & \mathbf{0} \end{pmatrix} + \begin{pmatrix} \mathbf{A}_\gamma \mathbf{Y}_\gamma^v \mathbf{A}_\gamma^T & -\mathbf{A}_\gamma \mathbf{Y}_\gamma^K \mathbf{A}_\gamma^T \\ \mathbf{A}_\gamma \mathbf{Y}_\gamma^K \mathbf{A}_\gamma^T & \mathbf{A}_\gamma \mathbf{Y}_\gamma^v \mathbf{A}_\gamma^T \end{pmatrix}, \tag{A.3.28}$$

where $\tilde{\mathbf{Y}}_\beta = \begin{pmatrix} \mathbf{Y}_\beta^K & \mathbf{0} \\ \mathbf{0} & \mathbf{0} \end{pmatrix}$ holds and the $\mathbf{0}$s denote zero matrices of appropriate size.

According to the discussion in Sect. A.2, in (A.3.28) the elements of the matrix that is the first term in the sum are of several orders of magnitude smaller than the elements of the matrix in the second term. In the case of $N_c < N$ the first matrix is obviously singular. On the basis of Theorem 8, it can be seen that the matrix in the second term of the sum is singular. Therefore, the matrix of the system of equations is nearly singular (see [26] concerning this notion), which causes severe difficulties in the numerical solution of the system of equations.

Therefore, we modify the system of linear equations (A.3.23) in the following way: we consider one of the components of the potential vector $\mathbf{u}$ as fixed and delete the equation with a serial number corresponding to the fixed component. The matrix of the new system of equations can be obtained by deleting the corresponding row and column in matrix $\hat{\mathbf{Y}}$. This means deleting two rows and two columns with corresponding serial numbers in the matrix $\tilde{\mathbf{Y}}$ of the real system of equations. This implies that the role of $\mathbf{A}_\gamma$ in the second term of (A.3.28) is now played by the corresponding reduced node-edge incidence matrix $\mathbf{A}_\gamma^r$ of the graph $(\mathcal{N}_\gamma, \mathscr{A}_\gamma^D)$. According to Theorem 8, the rows of $\mathbf{A}_\gamma^r$ are linearly independent and the diagonal elements of $\mathbf{Y}_\gamma^v$ are positive. Consequently, it is easy to see that $\mathbf{A}_\gamma^r \mathbf{Y}_\gamma^v (\mathbf{A}_\gamma^r)^T$ is a positive definite matrix.

Both the matrix of the modified system of equations and the second term matrix in its decomposition, carried out analogously to the case of (A.3.28), are nonsingular, whichever row has been deleted. On the basis of our discussions so far, this fact easily follows from the following theorem.

Theorem 1 *Let us consider a matrix* $\mathbf{D}$ *of size* $2p \times 2p$ *with the following structure:*

$$\mathbf{D} = \begin{pmatrix} \mathbf{D}_1 & \mathbf{D}_3 \\ -\mathbf{D}_3 & \mathbf{D}_2 \end{pmatrix}, \tag{A.3.29}$$

where $\mathbf{D}_1$ *and* $\mathbf{D}_2$ *are symmetric positive-definite matrices of size* $p \times p$ *and* $\mathbf{D}_3$ *is a symmetric matrix of size* $p \times p$. *Then* $\mathbf{D}$ *is nonsingular.*

Proof. Let $\mathbf{x} \in \mathbb{R}^{2p}$ be a vector for which $\mathbf{Dx} = \mathbf{0}$ holds, and let us partition x as

$$\mathbf{x} = \begin{pmatrix} \mathbf{x}_1 \\ \mathbf{x}_2 \end{pmatrix}, \mathbf{x}_1, \mathbf{x}_2 \in \mathbb{R}^p.$$ Because of our assumptions we have

$$\mathbf{x}^T \mathbf{Dx} = \mathbf{x}_1^T \mathbf{D}_1 \mathbf{x}_1 + \mathbf{x}_2^T \mathbf{D}_2 \mathbf{x}_2 = 0. \tag{A.3.30}$$

This relation implies that $\mathbf{x} = \mathbf{0}$ holds because of the positive definiteness of $\mathbf{D}_1$ and $\mathbf{D}_2$. Therefore, the columns of $\mathbf{D}$ are linearly independent, i.e., $\mathbf{D}$ is nonsingular and the theorem is proved.   □

Furthermore, in the (A.3.28)-type decomposition when carried out for the matrix of the modified system of equations, the second term is generally a well-conditioned matrix. Therefore, taking into account the difference involving orders of magnitude between the elements of the two matrices in the sum, the matrix of the modified system of equations will itself also be well conditioned.

Definition The node of the network $(\mathcal{N}, \mathscr{A}_\beta^D \cup \mathscr{A}_\gamma^D)$ for which the complex potential is considered as fixed is called the *reference node of the network*.

The preceding result can also be formulated in the following manner: if the current injections are prescribed for the nodes of the network $(\mathcal{N}, \mathscr{A}_\beta^D \cup \mathscr{A}_\gamma^D)$ other than the reference node, the modified system of equations derived from relation (A.3.23) uniquely determines the value of complex potentials of the nodes other than the reference node, and the system of equations is also acceptable from a numerical point of view. In relation (A.3.23), the current injection in the row corresponding to the reference node is expressed explicitly in terms of potentials at the other nodes; therefore, the current injection at the reference node can be computed via substitution of the solution after having solved the system of equations. These characteristics are unrelated to the fact of what node plays the role of the reference node.

The potential distribution depends on the choice of the reference point and, naturally, on the complex potential of the reference node.

The reference node of a transmission network cannot be chosen arbitrarily. On the basis of the preceding discussion, we see that the current injection at the reference node cannot be prescribed; it is uniquely determined by the potential distribution. Therefore, only those nodes can operate as reference nodes where the current injection is controllable, i.e., to which also power plants are connected with a sufficiently high capacity that can provide injection at the level of actual demand. Furthermore, the power plant that is attached to the reference node generally has an important role in system control (e.g., in frequency control) as well. This subject is beyond the scope of this book. The interested reader may consult [17] for further information.

In the case of the transmission network of electric power systems, the actually considered physical quantity is the electric power, instead of currents, with respect to generation, consumption, and transmission. Therefore, relation (A.3.23) must be reformulated in terms of powers instead of currents, which results in a loss of linearity of the relevant system of equations.

Let $\mathbf{s} \in \mathbb{C}^N$ denote the vector of powers attached to the branches. Then, on the basis of (A.1.17), the components of $\mathbf{s}$ can be computed as follows:

$$s_p = u_p i_p^*, \qquad p = 1, \ldots, N. \tag{A.3.31}$$

Utilizing this relation and relation (A.3.14), (A.3.23) can be formulated as

$$u_p^* \sum_{q=1}^{N} \hat{Y}_{pq} u_q = -s_p^*, \qquad p = 1, \ldots, N. \tag{A.3.32}$$

Regarding the components of the potential vector $\mathbf{u}$ as unknowns, we have arrived at a nonlinear system of equations. Taking this as our starting point, we formulate the power flow (so-called load flow) problem of transmission networks.

For this purpose we will introduce a new system of notations for electric quantities in order to follow the generally accepted notations in the engineering literature. This system of notations is used in the remaining parts of the appendix.

For the sake of simplicity of presentation, it is assumed that a β-type capacitive branch is connected to every node, with the prescription $C_i = 0$, $i = N_c + 1, \ldots, N$.

In the sequel we will concentrate our analysis on the network $(\mathcal{N}, \mathscr{A}_\beta^D \cup \mathscr{A}_\gamma^D)$, which has N nodes and $N_c + M$ edges. We will work with node potentials, and their vector will be denoted by $\mathbf{U} \in \mathbb{C}^N$.

Let $U_k = V_k e^{j\theta_k}$, $k = 1, \ldots, N$, where V_k is the absolute value of the potential and θ_k is its phase angle. The vectors with components V_k and θ_k, $k = 1, \ldots, N$, are denoted by $\mathbf{V}$ and θ, respectively.

The representation with *Cartesian coordinates* will be needed too, and for this lowercase letters will be used in the system of notations to be in accordance with the system of notations of the scheduling model. Therefore, let $\mathbf{U} = \mathbf{v} + j\mathbf{w}$, $\mathbf{v} \in \mathbb{R}^N$, $\mathbf{w} \in \mathbb{R}^N$.

As discussed previously, the node potentials of the model can be interpreted as the edge voltages of α-type edges. In the sequel, node potentials will be called *node voltages* or simply *voltages*. Regarding their physical content, they are voltages of the busbars with respect to the grounded node, and they are important measured parameters of the system.

Let $\mathbf{S} \in \mathbb{C}^N$ denote the vector formed by (-1) times the powers associated with the edges of a tree whose components, in accordance with our previous discussion, can be interpreted as a power injection at the nodes of the graph $(\mathcal{N}, \mathscr{A}_\beta^D \cup \mathscr{A}_\gamma^D)$.

Let $\mathbf{S} = \mathbf{P} + j\mathbf{Q}$, $\mathbf{P} \in \mathbb{R}^N$, $\mathbf{Q} \in \mathbb{R}^N$, where the components of $\mathbf{P}$ are the active powers and the components of $\mathbf{Q}$ are the reactive powers. Consumption is interpreted as negative injection.

Let $\mathbf{I} \in \mathbb{C}^M$ denote the vector of edge currents belonging to γ-type edges of the network, and let $\mathbf{I} = \mathbf{I}^P + j\mathbf{I}^Q$, where the components of $\mathbf{I}^P \in \mathbb{R}^M$ are called *real currents* and the components of $\mathbf{I}^Q \in \mathbb{R}^M$ are called *imaginary* or *reactive currents*.

Subsequently, it is assumed that the edges of $(\mathcal{N}, \mathscr{A}_\beta^D \cup \mathscr{A}_\gamma^D)$ are numbered from 1 to $N_c + M$ in such a way that the numbering starts with the β-type edges.

In accordance with the discussion in Sect. A.2, the edge admittances of the network are as follows:

$$Y_{N+1,i} = j\omega C_i, \qquad\qquad i = 1, \ldots, N_c, \qquad\qquad (A.3.33)$$

and

$$Y_{i,l} = \frac{1}{R_{il} + j\omega L_{il}}, \qquad (i,l) \in \mathscr{A}_\gamma^D. \qquad\qquad (A.3.34)$$

The following notations are introduced:

$$\left. \begin{aligned} X_{il} &= \omega L_{il} \\ G_{il} &= \frac{R_{il}}{R_{il}^2 + X_{il}^2} \\ B_{il} &= \frac{X_{il}}{R_{il}^2 + X_{il}^2} \end{aligned} \right\} \quad (i,l) \in \mathscr{A}_\gamma^D. \qquad\qquad (A.3.35)$$

Let $Y_{li} = Y_{il}$, $X_{li} = X_{il}$, $G_{li} = G_{il}$, and $B_{li} = B_{il}$ for all $(i,l) \in \mathscr{A}_\gamma^D$. Therefore, we get the following relation for the edge admittances of γ-type edges:

$$Y_{il} = G_{il} - jB_{il}, \qquad (i,l) \in \mathscr{A}_\gamma. \tag{A.3.36}$$

The diagonal elements of matrix $\mathbf{Y}_{KK}$ on the right-hand side of relation (A.3.14) are the previously specified edge admittances.

It is easy to see that matrix $\hat{\mathbf{Y}}$, defined by relation (A.3.14), can be specified directly in the following form:

$$\hat{Y}_{ik} = \begin{cases} \displaystyle\sum_{l \in J(i)} Y_{il} + Y_{n+1,i}, & \text{if } i = k, \\ -Y_{ik}, & \text{if } i \neq k, k \in J(i), \\ 0, & \text{otherwise.} \end{cases} \tag{A.3.37}$$

In the representation (A.3.37) and later in the text as well, the notation $J(i)$ introduced in Sect. A.1 refers to the graph $(\mathscr{N}_\gamma, \mathscr{A}_\gamma^D)$.

The real and imaginary parts of the elements of matrix $\hat{\mathbf{Y}}$ will also be needed; therefore, the following notations are introduced:

$$\hat{Y}_{ik} = \hat{G}_{ik} + j\hat{B}_{ik}; \qquad \hat{G}_{ik} \text{ and } \hat{B}_{ik} \text{ are real}, i,k = 1,\dots,N. \tag{A.3.38}$$

Following some minor reformulations, in our new system of notations, relations (A.3.32) take the following form:

$$U_i \sum_{k=1}^{N} \hat{Y}_{ik}^* U_k^* = S_i, \qquad i = 1,\dots,N. \tag{A.3.39}$$

By virtue of some simple rearrangements and utilizing representation (A.3.37), the following relation is obtained:

$$|U_i|^2 Y_{n+1,i}^* + \sum_{k \in J(i)} U_i(U_i - U_k)^* Y_{ik}^* = S_i, \qquad i = 1,\dots,N. \tag{A.3.40}$$

Utilizing (A.3.38) and taking the polar coordinate form of the voltages, the equivalent real relations corresponding to equations (A.3.39) are as follows:

$$V_i \sum_{k=1}^{N} V_k[\hat{G}_{ik}\cos(\theta_i - \theta_k) + \hat{B}_{ik}\sin(\theta_i - \theta_k)] = P_i,$$

$$V_i \sum_{k=1}^{N} V_k[\hat{G}_{ik}\sin(\theta_i - \theta_k) - \hat{B}_{ik}\cos(\theta_i - \theta_k)] = Q_i, \qquad i = 1,\dots,N. \tag{A.3.41}$$

In the case of the *Cartesian coordinates*, the equivalent real relations are as follows:

$$\sum_{k=1}^{N}(\hat{G}_{ik}[v_i v_k + w_i w_k] + \hat{B}_{ik}[v_k w_i - v_i w_k]) = P_i,$$

$$\sum_{k=1}^{N}(\hat{G}_{ik}[v_k w_i - v_i w_k] - \hat{B}_{ik}[v_i v_k + w_i w_k]) = Q_i, \qquad i = 1,\dots,N. \quad \text{(A.3.42)}$$

Subsequently, let us denote by $\Phi_i(\mathbf{V},\theta)$ and $\Psi_i(\mathbf{V},\theta)$ the functions on the left-hand side of the equations in (A.3.41), corresponding to active and reactive power, respectively. The functions $f_i(\mathbf{v},\mathbf{w})$ and $g_i(\mathbf{v},\mathbf{w})$ are defined analogously with respect to relations (A.3.42).

With these notations, (A.3.41) and (A.3.42) take the following form:

$$\Phi_i(\mathbf{V},\theta) = P_i,$$

$$\Psi_i(\mathbf{V},\theta) = Q_i, \qquad i = 1,\dots,N, \qquad\qquad \text{(A.3.43)}$$

$$f_i(\mathbf{v},\mathbf{w}) = P_i,$$

$$g_i(\mathbf{v},\mathbf{w}) = Q_i, \qquad i = 1,\dots,N. \qquad\qquad \text{(A.3.44)}$$

The complex relations (A.3.39) and the equivalent real relations (A.3.43) and (A.3.44) describe the connection between power injections at the nodes and the node potentials. In the sequel, the real relations will be discussed. They are nonlinear, with $2N$ power type and $2N$ voltage type variables, connected by altogether $2N$ relations.

There are two relations and four variables for each node. If two variables are fixed and two are considered unknown at each node, then the number of the equations in the system of equations will be equal to the number of the unknowns.

We present a brief discussion on the question of which variables may be fixed at the various nodes.

At the reference node of the network the complex voltage will be fixed in such a way that in the polar representation the phase angle will be set to 0. The complex power injection is considered unknown. For the sake of simplicity let us assume that the reference node has a serial number of 1.

At the nodes corresponding to the consumer busbars, the complex power consumption is given and will be fixed, whereas the complex voltage is considered unknown.

The situation is the same at those nodes where the adjoining power plant generators are adjusted for fixed reactive power injection.

Compared to the preceding two types of fixing, those nodes where the active power injection and the absolute value of voltage are fixed can be considered an intermediate case. This fixing presupposes that, on the one hand, using devices connected to the node the aforementioned quantities can be adjusted, while on the other hand, reactive power injection, as prescribed by (A.3.43), takes place. Nodes having connections to a power plant or to a synchronous condenser belong to this type. In the case of power plants, this type of fixing is justified as follows. Active power generation can be controlled by the overall power plant control. The voltage of synchronous generators can be adjusted by controlling the current (excitation) of their rotor, and through this the absolute value of the voltage of the power plant busbar with respect to to the grounded node can also be controlled. The case is similar with respect to the synchronous condenser because it is also a synchronous engine. These nodes are called *voltage-preserving nodes* or *controllable sources of reactive power*. For the sake of simplicity we assume that the serial numbers of this type of node are $2, \ldots, L$.

On the basis of the preceding discussion, the nodes can be classified as follows:

(a) Reference node: V_1, $\theta_1 = 0$ are fixed; the unknowns are P_1 and Q_1.
(b) (P, Q) nodes: P_k and Q_k are fixed; V_k and θ_k are unknowns, $k = L+1, \ldots N$. Most (80–90 %) of network nodes are of this type.
(c) (P, V) nodes: P_k and V_k are fixed, Q_k and θ_k are unknowns, $k = 1, \ldots, L$.

The basic problem of load flow analysis can now be formulated as follows. The system of equations (A.3.43) respectively (A.3.44) should be solved by taking into account that, depending on the type of the nodes, the values of the adequate variables are fixed. The system of equations obtained in this way consists of $2N$ equations and has $2N$ unknowns.

In this system of equations those equations corresponding to the reference node and those corresponding to the reactive power relations for the (P, V) nodes can be disregarded for the following reason. In these equations one of the unknowns, appearing only in that specific equation, is expressed by the equation as a function of the other unknowns. Consequently, its value can be calculated by substitution after the values of the remaining unknowns are determined.

Thus we have obtained a system of equations of the network calculation problem for high-voltage transmission networks called load flow equations.

Let V_k^0, $k = 1, \ldots, L$ denote the absolute values of the fixed voltages.

In the case of polar coordinates, the following system of equations arises:

$$P_i - \Phi_i(\mathbf{V}, \theta) = 0, \qquad i = 2, \ldots, N,$$
$$Q_i - \Psi_i(\mathbf{V}, \theta) = 0, \qquad i = L+1, \ldots, N,$$
$$V_k = V_k^0, \qquad k = 1, \ldots, L,$$
$$\theta_1 = 0. \qquad\qquad\qquad\qquad\qquad (A.3.45)$$

Once the previously discussed variables are fixed, the unknowns are V_i, $i = L+1, \ldots, N$; $\theta_i = 2, \ldots, N$. The number of unknowns is $2N - L - 1$, which equals the number of equations.

In the case of the variant with *Cartesian coordinates* the system of equations is as follows:

$$P_i - f_i(\mathbf{v}, \mathbf{w}) = 0, \qquad i = 2, \dots, N,$$

$$Q_i - g_i(\mathbf{v}, \mathbf{w}) = 0, \qquad i = L+1, \dots, N,$$

$$v_k^2 + w_k^2 = (V_k^0)^2, \qquad k = 1, \dots, L,$$

$$v_1 = V_1^0,$$

$$w_1 = 0. \tag{A.3.46}$$

Once the previously discussed variables are fixed, the unknowns are v_i, $i = 2, \dots, N$, w_i, $i = 2, \dots, N$. The number of unknowns is $2N - 2$, which equals the number of equations.

At the nodes connected to controllable sources of reactive power, bounds depending on the characteristics of the adjoining devices must be satisfied with respect to reactive power generation respectively consumption. If at a (P, V) node (A.3.43) can only be fulfilled by violating the bounds, this node should be changed to a (P, Q) type of node. The situation is analogous for (P, Q) nodes with respect to the voltage. This problem is customarily solved as follows. The load flow problem (A.3.43) is initially formulated with a fixed qualification of the nodes, and the solution algorithm is modified such that during the iterations the qualification of the nodes is changed if necessary. A detailed discussion of this topic is beyond the scope of this book; the interested reader may consult [17], for example.

A.4   Power Flow: Stott's Method for Solving the Load Flow Problem

Instead of currents, our model is based on power injected at the nodes, and for this reason the notion of power "flowing out" from the node to the branch will be used. Having defined the notion of a power flow, we can summarize the main characteristics of a transmission network of electric power systems based on the algorithm to be presented. Then comes a description of *Stott's method*, and finally some further relations are presented related to active power flow and transmission losses.

The starting point for the definition of *power flow* is relation (A.3.40). The following notations are introduced:

$$S_{ik}^\gamma = U_i(U_i - U_k)^* Y_{ik}^* \quad \text{if} \quad (i,k) \in \mathscr{A}_\gamma^D \quad \text{or} \quad (k,i) \in \mathscr{A}_\gamma^D, \tag{A.4.1}$$

$$S_{n+1,k}^\beta = |U_i|^2 Y_{n+1,i}^*, \qquad i = 1, \dots, N. \tag{A.4.2}$$

With our notations (A.3.40) takes the following form:

$$S_{n+1,i}^\beta + \sum_{k \in J(i)} S_{ik}^\gamma = S_i, \qquad i = 1, \dots, N. \tag{A.4.3}$$

Here S_i is the power injected at node i, and if the quantities S_{ik}^{γ}, S_{ik}^{β} are considered as powers flowing out from node i onto the adjoining branches, then (A.4.3), related to the power flow, is formally analogous to *Kirchhoff's current law*. It is easy to see that as physical quantities, S_{ik}^{β} and S_{ik}^{γ} represent power.

The quantities S_{ik}^{β} and S_{ik}^{γ} determined by (A.4.1) and (A.4.2) are considered jointly as a power flow on the network $(\mathcal{N}, \mathcal{A}_{\beta}^{D} \cup \mathcal{A}_{\gamma}^{D})$. The corresponding real and imaginary parts are called *active* and *reactive power flows*.

In this way, two power values are assigned to every $(i,k) \in \mathcal{A}_{\gamma}^{D}$ edge, and in this relation they are flowing out from the end nodes of an edge onto the edge, the latter being considered undirected in this relation. It is important to note that $S_{ik}^{\gamma} \neq -S_{ki}^{\gamma}$ holds, meaning that the power flowing out from node i is not equal to the power flowing in at the other node. Indeed, the value of the power flowing in at node k can be considered as (-1) times the amount of power flowing out into the edge (i,k). Therefore, if S_{ik}^{ν} denotes the difference between the two values of power, the following relation holds:

$$S_{ik}^{\nu} = S_{ik}^{\gamma} + S_{ki}^{\gamma} = |U_i - U_k|^2 Y_{ik}^*, \qquad (i,k) \in \mathcal{A}_{\gamma}^{D}. \tag{A.4.4}$$

Considering the physical content of the quantities S_{ik}^{ν}, they represent transmission losses at the edges of the network $(\mathcal{N}_{\gamma}, \mathcal{A}_{\gamma}^{D})$.

The following relation holds for the overall transmission loss S^{ν} as defined in Sect. A.3:

$$S^{\nu} = \sum_{k=1}^{N} |U_k|^2 Y_{n+1,k}^* + \sum_{(i,k) \in \mathcal{A}_{\gamma}^{D}} |U_i - U_k|^2 Y_{ik}^*. \tag{A.4.5}$$

In fact, this relation can easily be derived from Eq. (A.3.22) by utilizing (A.3.14) and (A.3.37).

Next, some of the characteristics of the transmission network are discussed, with the aim of preparing the construction of the algorithm.

Some branches of a transmission network represent transformers. There is a considerable potential difference among the endpoints of these branches. The stability of the algorithms aimed at solving the network calculation problem requires a proper standardization of the voltages at the nodes, which has further advantages as well [17]. Two types of procedures are used.

The first procedure is called the *reduction to a mutual voltage base*. First a voltage value is fixed, then all the voltages at the nodes are standardized for an approximate value. The fixed value is called the voltage constant. To each of the voltage intervals [100 kV, 200 kV), [200 kV, 400 kV), [400 kV, 700 kV), [700 kV, 900 kV) is assigned a nominal voltage value. The transformation is as follows:

$$V_{\text{new}} = V_{\text{old}} \frac{\text{voltage constant}}{V_{\text{nominal}}}. \tag{A.4.6}$$

To obtain the same values of power as those prior to the transformation, the remaining electric quantities must be transformed adequately.

The second procedure involves the introduction of relative units (r.u.). Quite frequently this means that the voltages at the nodes are standardized to obtain new values around the value 1. The transformation is as follows:

$$V_{\text{r.u.}} = \frac{V_{\text{old}}}{V_{\text{nominal}}}. \tag{A.4.7}$$

This procedure results in values without physical dimension, and, in agreement with the remark concerning the previous procedure, the remaining electric quantities must be transformed accordingly.

Next, those characteristics of the transmission network are summarized that play an important role in the construction of algorithms for solving network calculation problems.

(a) In the case of high-voltage electric networks, the inductive reactance X_{ik} of transmission lines and cables are generally much higher than their ohmic resistance R_{ik}. The ratio is one order of magnitude higher in the case of transformers. This implies that

$$G_{ik} \ll B_{ik}, \text{ for every } (i,k) \in \mathscr{A}_{\gamma}^{D}.$$

(b) During stable system operation, the phase angles θ_k with respect to the reference point are small, and $|\theta_k| < 20°$, $k = 1,\ldots,N$, holds. Phase differences related to individual branches are small, too: $|\theta_i - \theta_k| < 10°$ for all $(i,k) \in \mathscr{A}_{\gamma}^{D}$.

(c) In normal operating mode the deviation between the standardized voltages is smaller than 15 %.

It is an empirical fact that in the relationship between active power flow and voltage distribution [described by the relations concerning P_i in the system of equations (A.3.43)], the phase angles of voltages play the dominant part. Experience is similar with respect to the reactive power flow and the absolute values of voltages. As an explanation for this phenomenon, some heuristic reasons are given next; see [37].

It is first assumed that $|U_1| = |U_2|$ holds for the complex potentials U_1 and U_2 at the endpoints of the branch $(1,2)$. Let the current flowing in the branch be I (Fig. A.8).

In accordance with the preceding characteristics, δ is small and α is approximately 90° due to $G_{12} \ll B_{12}$. Therefore, I and U_1 (respectively U_2) are almost in the same phase. For this reason the active component of the power is large compared to the reactive power (in absolute values).

Next it is assumed that the potentials are in the same phase and their sizes differ (Fig. A.9).

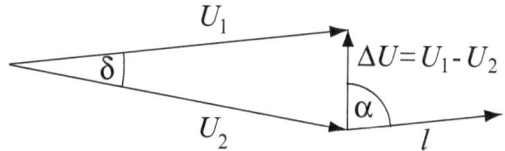

Fig. A.8 Case where potentials at endpoints of a branch have the same magnitude

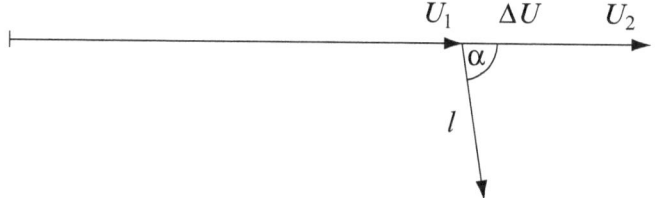

Fig. A.9 Case where potentials at endpoints of a branch are in the same phase

Then the phase angles of I and U_1 (respectively U_2) differ by approximately $90°$; therefore, the reactive component of the power is large compared to the active component.

For further analysis of the phenomenon, the system of equations (A.3.41) is decomposed into systems of equations corresponding to real and imaginary parts, and some simple reformulations are carried out. We introduce the following notations:

$$\delta_{ik} = \theta_i - \theta_k,$$

$$A(V_i, V_k, \delta_{ik}) = V_i^2 - V_i V_k \cos \delta_{ik},$$

$$B(V_i, V_k, \delta_{ik}) = V_i V_k \sin \delta_{ik}. \tag{A.4.8}$$

With these notations the system of equations takes the following form:

$$P_i = \sum_{k \in J(i)} [G_{ik} A(V_i, V_k, \delta_{ik}) + B_{ik} B(V_i, V_k, \delta_{ik})],$$

$$Q_i = \sum_{k \in J(i)} [B_{ik} A(V_i, V_k, \delta_{ik}) - G_{ik} B(V_i, V_k, \delta_{ik})] - \omega C_i V_i^2, \qquad i = 1, \dots, N. \tag{A.4.9}$$

Next, the partial derivatives of $A(V_i, V_k, \delta_{ik})$ and $B(V_i, V_k, \delta_{ik})$ are computed:

$$\frac{\partial A}{\partial V_i} = 2V_i - V_k \cos \delta_{ik}, \qquad \frac{\partial A}{\partial V_k} = -V_i \cos \delta_{ik},$$

$$\frac{\partial A}{\partial \delta_{ik}} = V_i V_k \sin \delta_{ik},$$

$$\frac{\partial B}{\partial V_i} = V_k \sin \delta_{ik}, \qquad\qquad \frac{\partial B}{\partial V_k} = V_i \sin \delta_{ik},$$

$$\frac{\partial B}{\partial \delta_{ik}} = V_i V_k \cos \delta_{ik}. \qquad\qquad\qquad\qquad (A.4.10)$$

According to the previous discussion, δ_{ik} is small, and if standardization to relative units is chosen, the absolute value of voltages is close to 1. Then $\left|\dfrac{\partial A}{\partial V_i}\right|$ and $\left|\dfrac{\partial A}{\partial V_k}\right|$ are significantly higher (they are close to 1) than $\left|\dfrac{\partial A}{\partial \delta_{ik}}\right|$ (which can have a value of at most around $\sin 10°$), i.e., the function $A(V_i, V_k, \delta_{ik})$ is mainly sensitive to the absolute values of the potentials. Similarly, it can be seen that the function $B(V_i, V_k, \delta_{ik})$ in the first line is sensitive to phase-angle differences.

Taking into account that $B_{ik} \gg G_{ik}$ holds for all $(i,k) \in \mathcal{A}_\gamma^D$ edges, (A.4.9) implies that the active power flow is mainly sensitive to the phase-angle differences, while the reactive power flow is mainly sensitive with respect to the absolute values of the voltages.

Next we present *Stott's method* [69] for the solution of the network calculation problem (A.3.45), which utilizes the previously discussed characteristics of a transmission network. The algorithm is a variant of *Newton's method*. The reason it was chosen is that due to the previously discussed characteristics of a transmission network, it is fairly easy to provide a reasonably good starting point. One such starting solution is the so-called flat voltage start, where the complex voltage of all the (P, Q) nodes equals that of the reference node. The same thing holds regarding the phase angles at (P, V) nodes. The choice of *Newton's method* is also justified by its fast convergence.

Based on the characteristics of the power flow discussed earlier in this appendix, a decomposed version of *Newton's method* is constructed.

Let us assume that a starting pair of vectors $\mathbf{V}^{(0)}$, $\theta^{(0)}$ is given, where $V_1^{(0)}, \ldots, V_L^{(0)}$ are the values prescribed in (A.3.45). Then if before the $(n+1)$th iteration $(n \geq 1)$ $\mathbf{V}^{(n)}, \theta^{(n)}$ is given, the next iteration of the algorithm runs as follows.

Step 1 A *Newton correction step* is carried out starting with $\theta^{(n)}$ to solve the following system of equations:

$$P_i - \Phi_i(\mathbf{V}^{(n)}, \theta) = 0, \qquad i = 2,\ldots,N,$$

$$\theta_1 = \theta_1^{(0)}. \qquad\qquad\qquad\qquad (A.4.11)$$

Let $\theta^{(n+1)}$ be the vector obtained in this way.

Step 2 A *Newton correction step* is carried out where $\mathbf{V}^{(n)}$ is the starting point for the following system of equations:

$$Q_i - \Psi_i(\mathbf{V}, \theta^{(n+1)}) = 0, \qquad i = L+1, \ldots, N,$$

$$V_k = V_k^{(0)}, \qquad k = 1, \ldots, L. \qquad (A.4.12)$$

Let the vector computed in this way be $\mathbf{V}^{(n+1)}$.

Starting from this algorithmic framework, *Stott's method* [69] can be obtained by neglecting some terms in the *Jacobi matrix*, which is applied in the *Newton steps*, by utilizing the characteristics of the transmission network.

The elements of the *Jacobi matrix* belonging to the system of equations (A.4.11) are

$$\left.\frac{\partial(P_i - \Phi_i(\mathbf{V}^{(n)}, \theta))}{\partial \theta_k}\right|_{\theta=\theta^{(n)}} = -V_i^{(n)} V_k^{(n)} \Big[G_{ik} \sin(\theta_i^{(n)} - \theta_k^{(n)})$$

$$- B_{ik} \cos(\theta_i^{(n)} - \theta_k^{(n)}) \Big], \quad \text{if} \quad i \neq k,$$

$$\left.\frac{\partial(P_i - \Phi_i(\mathbf{V}^{(n)}, \theta))}{\partial \theta_i}\right|_{\theta=\theta^{(n)}} = \Psi_i(V^{(n)}, \theta^{(n)}) + B_{ii}(V_i^{(n)})^2,$$

$$i, k = 2, \ldots, N. \qquad (A.4.13)$$

The elements of the *Jacobi matrix* in the case of the system of equations (A.4.12) are

$$\left.\frac{\partial(Q_i - \Psi_i(\mathbf{V}, \theta^{(n+1)}))}{\partial V_k}\right|_{\mathbf{V}=\mathbf{V}^{(n)}} = -V_i^{(n)} \Big[G_{ik} \sin(\theta_i^{(n+1)} - \theta_k^{(n+1)})$$

$$- B_{ik} \cos(\theta_i^{(n+1)} - \theta_k^{(n+1)}) \Big], \quad \text{if} \quad i \neq k,$$

$$\left.\frac{\partial(Q_i - \Psi_i(\mathbf{V}, \theta^{(n+1)}))}{\partial V_i}\right|_{\mathbf{V}=\mathbf{V}^{(n)}} = \frac{1}{V_i^{(n)}} \Big[-\Psi_i(\mathbf{V}^{(n)}, \theta^{(n+1)}) + B_{ii}(V_i^{(n)})^2 \Big]$$

$$i, k = L+1, \ldots, N. \qquad (A.4.14)$$

The following relations, implied by our discussions in this section, are the basis for carrying out some deletions in the *Jacobi matrix*:

$$\cos(\theta_i - \theta_k) \sim 1, \qquad |G_{ik} \sin(\theta_i - \theta_k)| \ll B_{ik}, \qquad (A.4.15)$$

for all $(i,k) \in \mathscr{A}_\gamma^D$. Accordingly, in relations (A.4.13) respectively (A.4.14), the terms $G_{ik} \sin(\theta_i - \theta_k)$ are deleted and $\cos(\theta_i - \theta_k)$ is replaced by 1.

Computer experimentation conducted over the course of developing this method revealed that $|\Psi_i|$ could be deleted as compared to $B_{ii} V_i^2$.

We introduce some further notations. Let $\Delta\theta^{(n+1)} - \theta^{(n)}$; then $\Delta\theta_1 = 0$ holds; $\Delta \mathbf{V} = \mathbf{V}^{(n+1)} - \mathbf{V}^{(n)}$, implying $\Delta V_i = 0$, $i = 1,\dots,L$.

With our notations, following some deletions and rearrangements, the system of linear equations corresponding to the *Newton correction* takes the following form:

$$V_i^{(n)} \sum_{k=2}^{N} B_{ik} V_k^{(n)} \Delta\theta_k = \Phi_i(\mathbf{V}^{(n)}, \theta^{(n)}) - P_i, \qquad i = 2,\dots,N, \qquad (A.4.16)$$

$$V_i^{(n)} \sum_{k=L+1}^{N} B_{ik} \Delta V_k = \Psi_i(\mathbf{V}^n, \theta^{(n+1)}) - Q_i, \qquad i = L+1,\dots,N. \qquad (A.4.17)$$

We apply one more approximation; the values of voltage levels $V_k^{(n)}$, $k = 2,\dots,N$ in (A.4.16) are considered to be equal. This common value is 1 in the case of r.u. scaling, described in Sect. A.2. If the other scaling is taken into account, the common value is the voltage constant V_{SKAL}. In this latter case the system of equations for correction is as follows:

$$\sum_{k=2}^{N} B_{ik} \Delta\theta_k = \frac{\Phi_i(\mathbf{V}^{(n)}, \theta^{(n)}) - P_i}{V_i^{(n)} V_{\mathrm{SKAL}}}, \qquad i = 2,\dots,N,$$

$$\sum_{k=L+1}^{N} B_{ik} \Delta V_k = \frac{\Psi_i(\mathbf{V}^{(n)}, \theta^{(n+1)}) - Q_i}{V_i^{(n)}}, \qquad i = L+1,\dots,N. \qquad (A.4.18)$$

The corresponding *Newton steps* will be as follows:

$$\theta^{(n+1)} = \theta^{(n)} + \Delta\theta, \qquad \mathbf{V}^{(n+1)} = \mathbf{V}^{(n)} + \Delta\mathbf{V}. \qquad (A.4.19)$$

A great advantage of this method is that, during iterations, the matrices of the system of linear equations in the calculations of the corrections are equal. Therefore, if the *Crout elimination* is applied, it is sufficient to carry out triangularization just once; in subsequent iterations the solution can be obtained by simple back substitutions.

Let us now return to the discussion of the subject of power flow and transmission losses. In the model for optimizing the daily schedule we will need in the case of γ-type branches the absolute value of power, i.e., the apparent power, flowing out from node i onto the edge (i,k), which is defined by the equality (A.4.1). To emphasize the dependence on the voltages at the nodes, this power is denoted by $S_{ik}(v_i, v_k, w_i, w_k)$. Disregarding the superscript in Eq. (A.4.1), the following formula can be obtained from (A.4.1) for the square of the apparent power:

$$|S_{ik}(v_i, v_k, w_i, w_k)|^2 = (v_i^2 + w_i^2)[(v_i - v_k)^2 + (w_i - w_k)^2]|Y_{ik}|^2, \quad (i,k) \in \mathscr{A}_\gamma^D.$$

(A.4.20)

Here the voltages were considered according to their real and imaginary parts, i.e., *Cartesian coordinates* were applied. This choice is motivated by the fact that in the model for optimal daily scheduling the real and imaginary parts of the voltages act as variables. In the final part of the appendix, we will retain this representation throughout.

The following notations are introduced with respect to losses on the edges of the network $(\mathscr{N}_\gamma, \mathscr{A}_\gamma^D)$, defined by Eqs. (A.4.4):

$$S_{ik}^v = P_{ik}^v + jQ_{ik}^v, \qquad P_{ik}^v, Q_{ik}^v \text{ are real, } (i,k) \in \mathscr{A}_\gamma^D.$$

(A.4.21)

From this the following formula can be derived for the active power loss:

$$P_{ik}^v = G_{ik}|U_i - U_k|^2.$$

(A.4.22)

Because of the capacitive character of β-type branches, it is obvious that there is no active power loss along those edges.

Relation (A.4.22) was derived utilizing the π circuit as a substitute for transmission lines or transformers. A more accurate value can be obtained if the imaginary current generated by the capacitance of the transmission line is added to the imaginary part of the corresponding edge current, taking into account that the vertical parts of π represent one half of the capacitive reactance of the edge. In this way, an error is corrected that arises upon substitution by the π circuit and has its origin within the π element. In the procedure for building the physical model in Sect. A.2, the vertical parts of the π circuits, used to represent the transmission lines and cables, were replaced by a single edge based on impedances connected in parallel. For the previously described correction we need the original vertical parts. Let C_{ik} denote the capacity value attached to edge $(i,k) \in \mathscr{A}_\gamma^D$ in the equivalent connection, considered in Sect. A.2. If the branch represents a transformer, then let $C_{ik} = 0$. Since the vertical parts of π represent one half of the capacity of transmission lines or cables, the edge admittance attached to them is $\frac{1}{2}j\omega C_{ik}$.

It is well known that the following formula holds for the active power loss [17]:

$$P_{ik}^v = R_{ik}|I_{ik}|^2.$$

(A.4.23)

Therefore, to determine the loss, the complex edge current is needed. Applying *Ohm's law* gives

$$I_{ik} = \frac{U_i - U_k}{R_{ik} + jX_{ik}} + j\frac{1}{2}v_i\omega C_{ik}, \qquad (i,k) \in \mathscr{A}_\gamma^D.$$

(A.4.24)

If in Eq. (A.4.24) the *Cartesian coordinates* of the voltages are chosen, the edge current will be as follows:

$$I_{ik} = (v_i - v_k)G_{ik} + (w_i - w_k)B_{ik} + j\left[(w_i - w_k)G_{ik} - (v_i - v_k)B_{ik} + \frac{1}{2}v_i\omega C_{ik}\right].$$

$$(A.4.25)$$

From this the following relations can be deduced:

$$I_{ik}^P = -G_{ik}(v_k - v_i) - B_{ik}(w_k - w_i),\qquad\qquad (A.4.26)$$

$$I_{ki}^P = -I_{ik}^P,\qquad\qquad\qquad\qquad\qquad\qquad\qquad\quad (A.4.27)$$

$$I_{ik}^Q = B_{ik}(v_k - v_i) - G_{ik}(w_k - w_i) + \frac{1}{2}v_i\omega C_{ik},\qquad (A.4.28)$$

$$I_{ki}^Q = -B_{ik}(v_k - v_i) + G_{ik}(w_k - w_i) + \frac{1}{2}v_k\omega C_{ik}.\qquad (A.4.29)$$

By virtue of substitution in (A.4.23), the value of the corrected loss results:

$$P_{ik}^v = \frac{1}{2}R_{ik}\left[2(I_{ik}^P)^2 + (I_{ik}^Q)^2 + (I_{ki}^Q)^2\right].\qquad (A.4.30)$$

In the sequel, the active power loss arising on the branch $(i,k) \in \mathscr{A}_\gamma^D$ will be denoted by $P_{ik}^v(v_i, v_k, w_i, w_k)$ to emphasize its dependence on the voltage, for every branch $(i,k) \in \mathscr{A}_\gamma^D$.

Accordingly, the overall loss in the network will be denoted by $P^v(\mathbf{v}, \mathbf{w})$. Its value is the sum of losses on the various edges

$$P^v(\mathbf{v}, \mathbf{w}) = \sum_{(i,k)\in\mathscr{A}_\gamma^D} P_{ik}^v(v_i, v_k, w_i, w_k).\qquad (A.4.31)$$

Finally, some relations with respect to power flow are discussed. Let $T_{ik}(v_i, v_k, w_i, w_k)$ denote the active power flowing out from node i into the branch $(i,k) \in \mathscr{A}_\gamma^D$. It can be seen from Eq. (A.4.1) that its value is as follows:

$$T_{ik}(v_i, v_k, w_i, w_k) = G_{ik}[v_i(v_i - v_k) + w_i(w_i - w_k)] + B_{ik}[w_i v_k - w_k v_i].\quad (A.4.32)$$

If the active power flowing out from the starting point onto the edge (i,k) has been determined, then utilizing this and the loss arising on this edge, the value of the power flowing out from node k onto the edge can be determined. This power flowing out from node k is denoted by $T_{ki}(v_i, v_k, w_i, w_k)$. From the definition of the loss we conclude that

$$T_{ki}(v_i, v_k, w_i, w_k) = -T_{ik}(v_i, v_k, w_i, w_k) + P_{ik}^v(v_i, v_k, w_i, w_k)\qquad (A.4.33)$$

holds.

References

1. Benders, J. F. (1962). Partitioning procedures for solving mixed-variables programming problems. *Numerische Mathematik, 4*, 238–252.
2. Billington, R., & Sachdeva, M. S. (1971). Real and reactive power optimization by suboptimum techniques. *IEEE Transactions on Power Apparatus and Systems, PAS-90*(2), 950–956.
3. Boros, E. (1986). Analysis and short-term forecasting of electricity demand. *Zeitschrift für Angewandte Mathematik und Mechanik, 66*, T340–T342.
4. Carpentier, J. W. (1973). Differential injections method, a general method for secure and optimal load flows. In *Proceedings of 8th PICA Conference* (pp. 255–262). Minneapolis.
5. Carpentier, J., & Siroux, J. (1963). L'optimisation de la production à l'Électricité de France. *Bulletin de la Société Française des Électriciens*, Ser.B, *4*, 121–129.
6. Cohen, A. I., & Yoshimura, M. (1983). A branch-and-bound algorithm for unit commitment. *IEEE Transactions on Power Apparatus and Systems, PAS-102*(2), 444–450.
7. Cohn, N. (1961). *Control of generation and power flow on interconnected power systems*. New York: Wiley.
8. Dantzig, G. B. (1963). *Linear programming and extensions*. Princeton: Princeton University Press.
9. Day, J. T. (1971). Forecasting minimum production costs with linear programming. *IEEE Transactions on Power Apparatus and Systems, PAS-90*(2), 814–823.
10. Deák, I., Hoffer, J., Mayer, J., Németh, Á., Potecz, B., Prékopa, A., et al. (1981). Optimal daily scheduling of the electricity production in Hungary. In G. B. Dantzig, M. A. H. Dempster, & M. Kallio (Eds.), *Large scale linear programming. Proceedings of a IIASA Workshop* (pp. 923–960). IIASA, Laxenburg, Austria.
11. Deák, I., Hoffer, J., Mayer, J., Németh, Á., Potecz, B., Prékopa, A., et al. (1982). Optimal daily scheduling of the electricity production in Hungary. In G. Feichtinger & P. Kall (Eds.), *Operations research in progress* (pp. 103–114). Dordrecht: D. Reidel Publishing Company.
12. Deák, I., Hoffer, J., Mayer, J., Németh, Á., Potecz, B., Prékopa, A., et al. (1983). Large-scale mixed-variable model for short-term optimal scheduling of power generation with thermal power plants under network constraints. *Alkalmazott Matematikai Lapok, 9*, 221–337 (In Hungarian).
13. Dillon, T. S., Edwin, K. W., Kochs, H. D., Tend, R. J. (1978). Integer programming approach to the problem of optimal unit commitment with probabilistic reserve determination. *IEEE Transactions on Power Apparatus and Systems, PAS-97*(6), 2154–2166.
14. Dillon, T. S., Martin, R. W., & Sjelogren, D. (1980). Stochastic optimization and modelling of large hydrothermal systems for long-term regulation. *IPC Business Press Electrical Power and Energy Systems, 2*(1), 2–20.

A. Prékopa et al., *Scheduling of Power Generation*, Springer Series in Operations Research and Financial Engineering, DOI 10.1007/978-3-319-07815-1,
© Springer International Publishing Switzerland 2014

15. Dommel, H. W., & Tinney, W. F. (1968). Optimal power flow solutions. *IEEE Transactions on Power Apparatus and Systems, PAS-87*, 1866–1876.
16. Dopazzo, J. F., Klitin, O. A., Stagg, G. W., Watson, M. (1967). An optimization technique for real and reactive power allocation. *Proceedings of the IEE, 55*(11), 1877–1885.
17. Elgerd, O. I. (1971). *Electrical energy systems theory: An introduction*. New York: McGraw-Hill.
18. Erisman, A. M., Neves, K. W., & Dwarakanath, M. H. (1980). *Electric power problems: The mathematical challenge*. Philadelphia: SIAM.
19. Escudero, L. F. (1982). On maintenance scheduling of production units. *European Journal of Operational Research, 9*, 264–274.
20. Fan, N., Izraelevitz, D., Pan, F., Pardalos, P. M., & Wang, J. (2012). A mixed integer programming approach for optimal power grid intentional islanding. *Energy Systems, 3*, 77–93.
21. Frank, S., Steponavice, I., Rebennack, S. (2012). Optimal power flow: A bibliographic survey I. Formulations and deterministic methods. *Energy Systems, 3*, 221–258.
22. Frank, S., Steponavice, I., Rebennack, S. (2012). Optimal power flow: A bibliographic survey II. Non-deterministic and hybrid methods. *Energy Systems, 3*, 259–289.
23. Fu, Y., & Shahidehpour, M. (2007). Fast SCUC for large-scale power systems. *IEEE Transactions on Power Systems, 22*, 2144–2151.
24. Fu, Y., Shahidehpour, M., Li, Z. (2005). Security-constrained unit commitment with AC constraints. *IEEE Transactions on Power Systems, 20*, 1538–1550.
25. Ganguly, S., Sahoo, N. C., Das, D. (2013). Recent advances on power distribution system planning: A state-of-the-art survey. *Energy Systems, 4*, 165–193.
26. Gill, P. E., Murray, W., Wright, M. H. (1991). *Numerical linear algebra and optimization* (Vol. 1). Redwood City: Addison-Wesley.
27. Gross, C. A. (1979). *Power system analysis*. London: Wiley.
28. Hano, I., Tamura, Y., Narita, S., & Matsumomoto, K. (1969). Real time control of system voltage and reactive power. *IEEE Transactions on Power Apparatus and Systems, PAS-88*(10), 1544–1559.
29. Happ, H. H. (1977). Optimal power dispatch: A comprehensive survey. *IEEE Transactions on Power Apparatus and Systems, PAS-96*, 841–854.
30. Happ. H. H., Johnson, R. C., Wright, W. J. (1971). Large scale hydro thermal unit commitment: Method and results. *IEEE Transactions on Power Apparatus and Systems, PAS-90*(3), 1373–1384.
31. Happ, H. H., & Wirgau, K. A. (1978). Static and dynamic VAR compensation in system planning. *IEEE Transactions on Power Apparatus and Systems, PAS-97*(5), 1564–1578.
32. Harhammer, P. G. (1975). Economic dispatch using mixed integer programming. In *Proceedings of 5th Power System Computation Conference* (pp. 1–17). Cambridge, UK.
33. Harhammer, P. G. (1978). Economic operation of multiple reservoir systems. In *Proceedings of 6th Power System Computation Conference*. Darmstadt.
34. Harhammer, P. G. (1979). Economic operation of electric power systems. In A. Prékopa (Ed.), *Survey of mathematical programming* (Vol. 3, pp. 179–194). Budapest: Akadémiai Kiadó.
35. Hoffer, J. (1979). Benders' partitioning procedure completed by the examination of feasible solutions. *Alkalmazott Matematikai Lapok, 5*, 177–190 (In Hungarian).
36. Hoffer, J. (1981). Solution of decision dependent supply-production problems by using Benders' partitioning procedure. *Alkalmazott Matematikai Lapok, 7*, 73–82 (In Hungarian).
37. Jolissaint, Ch. H., Arvanitides, N. V., Luenberger, D. G. (1972). Decomposition of real and reactive power flows: A method suited to on-line applications. *IEEE Transactions on Power Apparatus and Systems, PAS-91*, 661–670.
38. Kéri, G., Molnár, M., Németh, Á., Potecz, B., Prékopa, A., Turchányi, P., et al. (1974). Optimization models for the daily scheduling of the electricity production in the Hungarian power system. In *Department of Operations Research, Computer and Automation Institute of the Hungarian Academy of Sciences*, Budapest (In Hungarian).
39. Kirchmayer, L. K. (1958). *Economic operation of power systems*. New York: Wiley.
40. Kirchmayer, L. K. (1959). *Economic control of interconnected systems*. New York: Wiley.

41. Klos, A. (1970). Algebraic model of electrical network. *IEEE Transactions on Power Apparatus and Systems, PAS-89*, 240–262.
42. Krumm, L. A. (1976). A generalization of Newton's method for the control of energy systems. *Izvestiya Akademii Nauk USSR, Energetika i Transport, 3*, 3–20 (In Russian).
43. Lamont, J. W., Lesso, W. G., Rantz, M. W. (1979). Daily fossil fuel management. In Power Industry Computer Applications Conference, PICA-79, Cleveland, Ohio, *IEEE Conference Proceedings* (pp. 228–235).
44. Lasdon, L. S. (1972). *Optimization for large systems*. New York: The Macmillan Company.
45. Lauer, G. S., Sandell, N. R., Bertsekas, D. P., Posbergh, T. A. (1982). Solution of large-scale optimal unit commitment problems. *IEEE Transactions on Power Apparatus and Systems, PAS-101*(1), 79–86.
46. Lausanne, B., & Veret, C. *Description de la nouvelle version du modèle ORESTE de choix des démarrages des groupes thermiques*. EDF-DER, Service Études de Réseaux HR, 32–0504.
47. Ma, H., Shahidehpour, S. M. (1999). Unit commitment with transmission security and voltage constraints. *IEEE Transactions on Power Systems, 14*, 757–764.
48. Mariani, E., & Di Perna, A. (1971). Programmazione giornaliera delle centrali idroelettriche a bacine e a serbatoio in un sistema di produzione misto. *L'Energia Elettrica, 7*, 427–448.
49. Martínez-Crespo, J., Usaola, J., Fernández, J. L. (2006). Security-constrained optimal generation scheduling in large-scale power systems. *IEEE Transactions on Power Systems, 21*, 321–332.
50. Mayer, J., & Prékopa, A. (1985). On the load flow problem of electric power systems. In Hj. Wacker (Ed.), *Applied optmization techniques in energy problems* (pp. 321–340). Stuttgart: B.G. Teubner.
51. Németh, Á., Nagy, I. (1981). Daily load forecasting in the Hungarian national control center. In *Proceedings of VIIth Power System Computation Conference* (pp. 520–525). Lausanne.
52. Nemhauser, G. L., Wolsey, L. A. (1999). *Integer and combinatorial optimization*. New York: Wiley-Interscience.
53. O'Neill, R. P., Hedman, K. W., Krall, E. A., Papavasiliou, A., Oren, S. S. (2010). Economic analysis of the N-1 reliable unit commitment and transmission switching problem using duality concepts. *Energy Systems, 1*, 165–195.
54. Padhy, N. P. (2004). Unit commitment: A bibliographical survey. *IEEE Transactions on Power Systems, 19*, 1196–1205.
55. Pang, C. K., Sheble, G. B., Albuyeh, F. (1981). Evaluation of dynamic programming based methods and multiple area representation for thermal unit commitments. *IEEE Transactions on Power Apparatus and Systems, PAS-100*(3), 1212–1218.
56. Peschon, J., Piercy, D. S., Tinney, W. F., & Tveit, O. J. (1968). Optimum control of reactive power flow. *IEEE Transactions on Power Apparatus and Systems, PAS-87*, 40–48.
57. Potecz, B. (1967). *A systemic view of intermittent operations in an electrical power system*. Doctoral Thesis, Technical University of Budapest (In Hungarian).
58. Potecz, B. (1973). Computer program for the optimal scheduling of production states. In *Proceedings of the Conference on Applications of Computer Science in the Electric Power Industry*. Pécs, Hungary (In Hungarian).
59. Prékopa, A. (1968). *Linear programming I*. Budapest: János Bolyai Matematical Society (In Hungarian).
60. Quazza, G. (1976). Highlights on technological trends in the on-line optimization of power system operation. In *Proceedings of IEE International Conference on On-Line Operation and Optimization of Transmission and Distribution Systems, 22–25 June*. London.
61. Sachdeva, M. S., Billington, R., Peterson, C. A. (1977). Representative bibliography on load forecasting. *IEEE Transactions on Power Apparatus and Systems, PAS-96*(2), 697–700.
62. Sasson, A. M., Aboytes, F., Cardenas, K., Gomez, F., Viloria, F. (1972). A comparison of power systems static optimization techniques. *ETZ-A Bd, 93*, 520–527.
63. Seshu, S., Reed, M. B. (1961). *Linear graphs and electrical networks*. London: Addison-Wesley.

64. Sgep (1981). *Système de gestion énergétique prévisionelle*. Électricité de France, Service des Mouvements d'Énergie Paris.
65. Shen, C. M., Laughton, M. A. (1970). Power system load scheduling with security constraints using dual linear programming. *Proceedings of IEE, 117*(11), 2117–2127.
66. Stagg, G. W., El-Abiad, A. H. (1968). *Computer methods in power system analysis*. New York: McGraw-Hill.
67. Stott, B. (1972). Decoupled Newton load flow. *IEEE Transactions on Power Apparatus and Systems, PAS-91*, 1955–1959.
68. Stott, B. (1974). Review of load flow calculation methods. *Proceedings of IEEE, 62*(7), 916–929.
69. Stott, B., & Alsaç, O. (1974). Fast decoupled load flow. *IEEE Transactions on Power Apparatus and Systems, PAS-93*, 859–869.
70. Vágó, I. (1985) *Graph theory: Application to the calculation of electrical networks*. Amsterdam: Elsevier.
71. Waight, J. G., Albuyeh, F., & Bose, A. (1981). Scheduling of generation and reserve margin using dynamic and linear programming. *IEEE Transactions on Power Apparatus and Systems, PAS-100*(5), 2226–2230.
72. Weedy, B. M. (1967). *Electric power systems*. London: Wiley.
73. Wu, L., & Shahidehpour, M. (2010). Accelerating the Benders decomposition for network-constrained unit commitment problems. *Energy Systems, 1*, 339–376.
74. Zhai, Q., Guan, X., Cheng, J., & Wu, H. (2010). Fast identification of inactive security constraints in SCUC problems. *IEEE Transactions on Power Systems, 25*, 1946–1954.

Index

A. Prékopa et al., *Scheduling of Power Generation*, Springer Series in Operations
Research and Financial Engineering, DOI 10.1007/978-3-319-07815-1,
© Springer International Publishing Switzerland 2014